VOLUME FORTY THREE

THE ENZYMES

Mesoporous Silica-based Nanomaterials and Biomedical Applications, Part A

VOLUME FORTY THREE

THE ENZYMES

Mesoporous Silica-based Nanomaterials and Biomedical Applications, Part A

Edited by

FUYUHIKO TAMANOI

Integrated Institute for Cell-Materials Science, Institute for Advanced Study, Kyoto University, Kyoto, Japan;
Department of Microbiology, Immunology and Molecular Genetics, University of California, Los Angeles, CA, United States

ELSEVIER

ACADEMIC PRESS

An imprint of Elsevier

Academic Press is an imprint of Elsevier
50 Hampshire Street, 5th Floor, Cambridge, MA 02139, United States
525 B Street, Suite 1650, San Diego, CA 92101, United States
The Boulevard, Langford Lane, Kidlington, Oxford OX5 1GB, United Kingdom
125 London Wall, London, EC2Y 5AS, United Kingdom

First edition 2018

ISBN: 978-0-12-815112-9
ISSN: 1874-6047

For information on all Academic Press publications
visit our website at https://www.elsevier.com/books-and-journals

Publisher: Zoe Kruze
Acquisition Editor: Sam Mahfoudh
Editorial Project Manager: Shellie Bryant
Production Project Manager: Abdulla Sait
Cover Designer: Greg Harris

Typeset by SPi Global, India

CONTENTS

CONTRIBUTORS

Alejandro Baeza
Dpto. Química Inorgánica y Bioinorgánica, Universidad Complutense de Madrid, Instituto de Investigación Sanitaria Hospital 12 de Octubre i+12; CIBER de Bioingeniería, Biomateriales y Nanomedicina, CIBER-BBN; Dpto. Materiales y Producción Aeroespacial, ETSI Aeronáutica y del Espacio, Universidad Politécnica de Madrid, Madrid, Spain

C. Jeffrey Brinker
Chemical and Biological Engineering; Center for Micro-Engineered Materials, Advanced Materials Laboratory, University of New Mexico, Albuquerque, NM, United States

Shunai Che
School of Chemical Science and Engineering, Tongji University; School of Chemistry and Chemical Engineering, State Key Laboratory of Metal Matrix Composites, Shanghai Jiao Tong University, Shanghai, China

Peilin Chen
Research Center for Applied Sciences, Academia Sinica, Taipei, Taiwan

Wei Chen
Department of Chemistry and Biochemistry, University of California, Los Angeles, CA, United States

Yi-Ping Chen
Graduate Institute of Nanomedicine and Medical Engineering, Taipei Medical University, Taipei, Taiwan

Chi-An Cheng
Department of Chemistry and Biochemistry, University of California, Los Angeles, CA, United States

Jonas G. Croissant
Chemical and Biological Engineering; Center for Micro-Engineered Materials, Advanced Materials Laboratory, University of New Mexico, Albuquerque, NM, United States

Tian Deng
Department of Chemistry and Biochemistry, University of California, Los Angeles, CA, United States

Jean-Olivier Durand
Institut Charles Gerhardt Montpellier, UMR-5253 CNRS-UM-ENSCM, Montpellier, France

Lu Han
School of Chemical Science and Engineering, Tongji University, Shanghai, China

Navnita Kumar
Department of Chemistry and Biochemistry, University of California, Los Angeles, CA, United States

Chiung Wen Kuo
Research Center for Applied Sciences, Academia Sinica, Taipei, Taiwan

Mika Lindén
Department of Inorganic Chemistry II, Ulm University, Ulm, Germany

Yanhang Ma
School of Physical Science and Technology, ShanghaiTech University, Pudong, Shanghai, China

Miguel Manzano
Dpto. Química Inorgánica y Bioinorgánica, Universidad Complutense de Madrid, Instituto de Investigación Sanitaria Hospital 12 de Octubre i+12; CIBER de Bioingeniería, Biomateriales y Nanomedicina, CIBER-BBN, Madrid, Spain

Chung Yuan Mou
Department of Chemistry, National Taiwan University, Taipei, Taiwan

Feby Wijaya Pratiwi
Nanoscience and Technology Program, Taiwan International Graduate Program; Research Center for Applied Sciences, Academia Sinica; Department of Chemistry, National Taiwan University, Taipei, Taiwan

Fuyuhiko Tamanoi
Integrated Institute for Cell-Materials Science, Institute for Advanced Study, Kyoto University, Kyoto, Japan; Department of Microbiology, Immunology and Molecular Genetics, University of California, Los Angeles, CA, United States

Osamu Terasaki
School of Physical Science, ShanghaiTech University, Pudong, Shanghai, China

Maria Vallet-Regi
Dpto. Química Inorgánica y Bioinorgánica, Universidad Complutense de Madrid, Instituto de Investigación Sanitaria Hospital 12 de Octubre i+12; CIBER de Bioingeniería, Biomateriales y Nanomedicina, CIBER-BBN; Department of Chemistry in Pharmaceutical Sciences, School of Pharmacy, Universidad Complutense de Madrid and Networking Research Center on Bioengineering, Biomaterials and Nanomedicine (CIBER-BBN), Madrid, Spain

Ruining Wang
Department of Chemistry and Biochemistry, University of California, Los Angeles, CA, United States

Si-Han Wu
Graduate Institute of Nanomedicine and Medical Engineering, Taipei Medical University, Taipei, Taiwan

Yaping Zhang
School of Physical Science and Technology, ShanghaiTech University, Pudong, Shanghai, China

Jeffrey I. Zink
Department of Chemistry and Biochemistry, University of California, Los Angeles, CA, United States

PREFACE

In the past years, we have expanded the scope of this series to include cancer biochemistry, cancer therapy, and diagnosis. This volume and the next volume (43 and 44) belong to this line of scientific inquiry. In this volume, we feature nanotherapy using mesoporous silica nanoparticles (MSNs). We discuss structure of this remarkable nanomaterials including structural characterization and potential to be used as a delivery vehicle for anticancer drugs. In particular, we discuss controlled release capability of MSNs that have significantly advanced the field of drug delivery technologies. Mechanized nanoparticles that respond to external stimuli such as light, ultrasound, and magnetic field are described. We also describe the use of MSNs for imaging. In addition, biodistribution and biodegradability of MSNs are discussed. The issue of biodistribution and biodegradation is important and this will be further expanded in the next volume.

We would like to thank all the authors for preparing their chapters despite their busy schedule. Without their timely preparation, we could not have put together this volume. We would also like to thank editorial help by Shellie Bryant of Elsevier.

FUYUHIKO TAMANOI
August 10, 2018
Kyoto and Los Angeles

CHAPTER ONE

Overview of Studies Regarding Mesoporous Silica Nanomaterials and Their Biomedical Application

Maria Vallet-Regi*, Fuyuhiko Tamanoi†,‡,1
*Department of Chemistry in Pharmaceutical Sciences, School of Pharmacy, Universidad Complutense de Madrid and Networking Research Center on Bioengineering, Biomaterials and Nanomedicine (CIBER-BBN), Madrid, Spain
†Integrated Institute for Cell-Materials Science, Institute for Advanced Study, Kyoto University, Kyoto, Japan
‡Department of Microbiology, Immunology and Molecular Genetics, University of California, Los Angeles, CA, United States
[1]Corresponding author: e-mail address: fuyut@microbio.ucla.edu

Contents

Abstract

Synthesis of mesoporous silica nanomaterials occurred in the late 1980s and early 1990s. Since then, we have seen dramatic increase in the number of publication dealing with this nanomaterial. In particular, there have been a large number of publications regarding biomedical application of mesoporous silica nanoparticles including their use in cancer therapy and diagnosis. Various surface modification strategies have been explored to enhance drug loading and to target them to tumor. In addition, the nanoparticles have been armed with controlled release capability. In this review, we point out some crucial issues regarding the material, which will be discussed in more detail in chapters of this volume.

The Enzymes, Volume 43
ISSN 1874-6047
https://doi.org/10.1016/bs.enz.2018.07.001

1. DRAMATIC INCREASE IN THE NUMBER OF PUBLICATION REGARDING MESOPOROUS SILICA NANOMATERIALS

Research in mesoporous materials is at the limelight since it gave rise to substantial amount of work in chemical synthesis, responsible for an enormous amount of exciting practical applications for the welfare of society. From catalysis to medical and nanotechnology applications, there are numerous new materials developed regarding mesoporous silica materials. As described below, the development of ordered mesoporous silica in the early 1990s was a major breakthrough, which has triggered research activities in hundreds of laboratories around the world.

The number of applications has expanded dramatically that encompass the fields of catalysis to nanomedicine. This is evident from the results of our survey of the Web of Knowledge; we searched literature for these materials in the 26 years since their inception. Fig. 1A shows the number of publications concerning this type of material per year. As can be seen, the number of publication starts picking up in late 1990s and explodes by the start of this century. In the last year, there were more than 11,000 publications. The total number of publications is 130,000 since 1991.

If we narrow the search to the combination of three categories (mesoporous silica nanoparticles, drug delivery and responsive), we find the results shown in Fig. 1B. The number of papers in this category starts picking up after 2005 as a variety of nanoparticles with controlled drug release features were developed. In the last year, 2500 papers were scored in this category.

2. SYNTHESIS OF MESOPOROUS SILICA MATERIALS: INITIAL DISCOVERY

Mesoporous silica nanomaterials were demonstrated to be an important class of material that has found many applications. Because of the presence of thousands of pores, clever utilization of these materials has been developed over the years (Figs. 2 and 3). In late 1980s and early 1990s, pioneering studies on the synthesis of mesoporous silica materials were reported. This involved the use of surfactants to generate mesopores during the sol–gel synthesis using a precursor such as TEOS. In 1988 and in 1990, a Waseda University group produced nanomaterials named KSW-1 and FSM-16 (folded sheets mesoporous materials) [1]. FSM-16 is composed of hexagonal channels

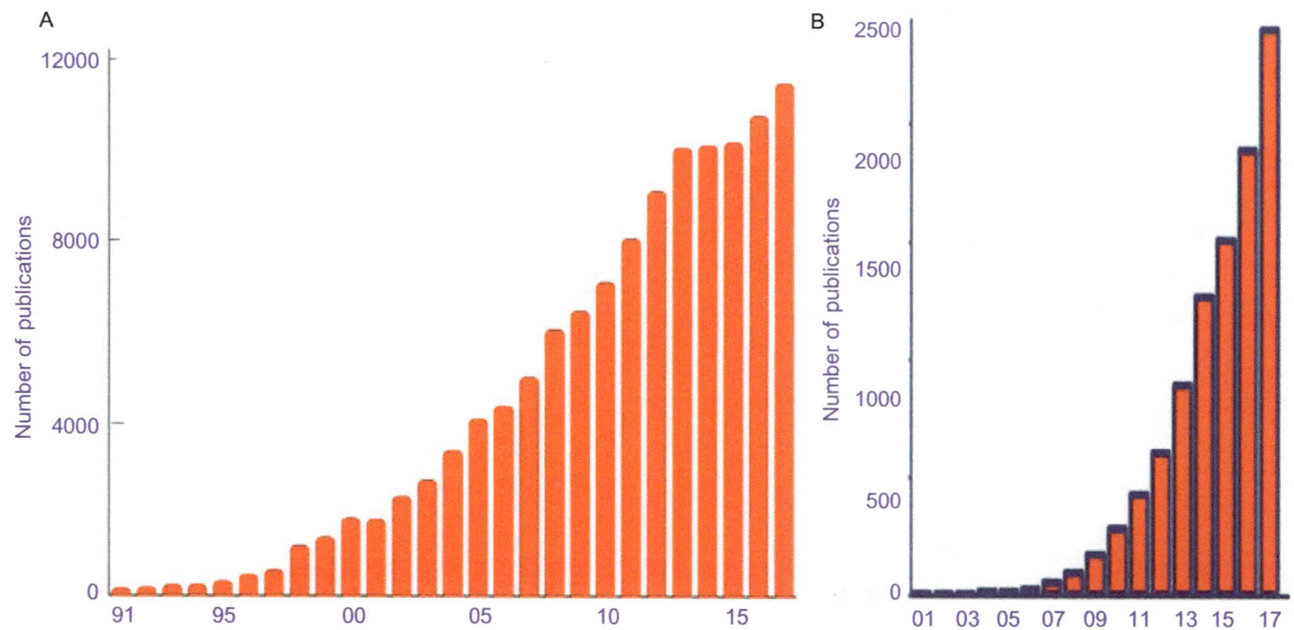

Fig. 1 Panel (A) shows the number of publications on mesoporous materials per year from 1991 to 2017. Panel (B) shows the number of publication of mesoporous silica nanoparticles with responsive drug delivery capability (from 2001 to 2017).

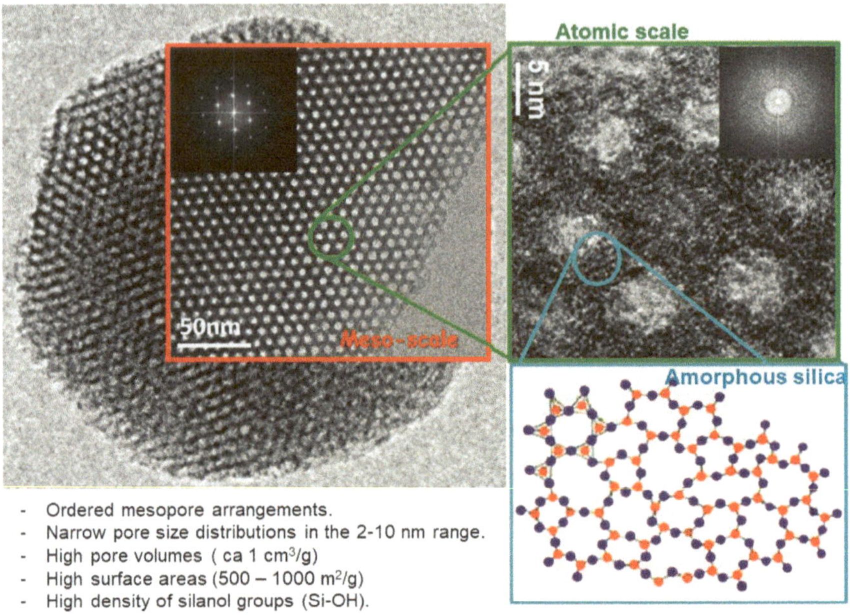

Fig. 2 Mesoporous silica nanoparticles observed at different scales by electron microscopy. Notable characteristics of these nanomaterials are listed.

and is synthesized by using quarternary ammonium surfactant as a template. Mobil corporation group reported synthesis of MCM-41 (Mobil Catalytic Material Number 41) nanomaterials in 1992 [2]. These contain uniform hexagonal pore structure. This was followed by the synthesis of similar porous materials, MCM-48 and MCM-50.

The establishment of the field of mesoporous material owes significantly to the physical demonstration of the presence of pore structure as well as materials encapsulated in the pores. This was possible by the use of electron microscopic methods. The work required a very difficult and complicated structural characterization of this family of materials, which are quite beam sensitive under the electron microscopy. This excellent characterization [3] has enabled understanding of the behavior of these materials, which was instrumental in projecting them to new and important applications.

Various other types of mesoporous silica nanomaterials have been developed. UC Santa Barbara group synthesized well-ordered hexagonal mesoporous silica structures SBA-15 (Santa Barbara Amorphous 15) by using amphiphilic block copolymers as organic structure directing agents [4]. Periodic mesoporous organosilica used bridged precursor to incorporate organic components into the framework of nanomaterials ([5] and reviewed in [6]).

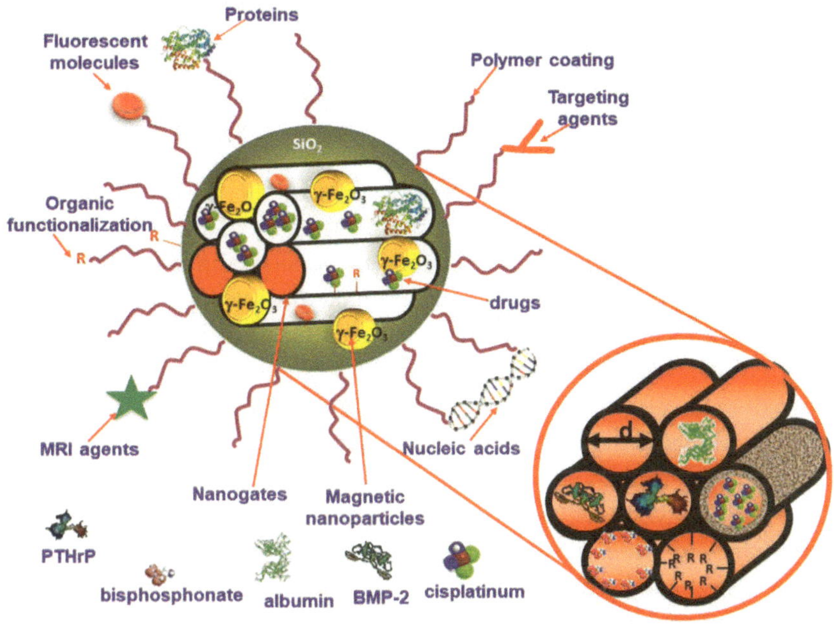

Fig. 3 Schematic representation of one mesoporous silica nanoparticle where it is possible to observe that both internal and external surface can be modified with various moieties that affect their drug loading and release behavior, tumor targeting capability and imaging effect.

Hollow mesoporous silica nanoparticles with a large hollow space have been produced and characterized for their use in biomedical application (reviewed in [7]).

Since the discovery of mesoporous materials in the early 1990s, many groups are involved in the synthesis and in the study, at a fundamental level, of mechanisms that are involved in the formation of mesoporous materials, finding new reaction pathways that have led to more complex nanostructures with a wider compositional range. To study the structural properties and porous characteristics of such materials, the development of characterization methods and, in particular, the use of electron crystallography has been, and still is, vital.

3. FUNCTIONALIZATION

One of the key features of mesoporous silica materials is the presence of large surface area. In fact, 1 g of MSN contains 500–1500 m^2 of surface area. This is quite remarkable, as the surface area provides a vast space to attach various chemical groups (Figs. 2 and 3). In many cases, it is necessary to chemically modify the mesoporous silica walls through the covalent attachment of functional groups. This functionalization process leads to hybrid mesoporous materials that can act as host matrices of a wide range of drugs via weak interactions.

The functionalization of the silica walls may be necessary for several reasons. In some cases, there are certain drugs with remarkable hydrophobic nature that do not exhibit any tendency to penetrate into the hydrophilic mesoporous silica. The functionalization with hydrophobic functional groups is a good alternative to promote loading of different hydrophobic drugs. This strategy is also employed to delay the release kinetics of certain drugs from mesoporous channels to the aqueous medium due to the decrease in the wettability degree of the material surface. On the other hand, there are other situations in which the pharmaceutical molecule can be confined into the mesoporous channels. However, higher loads and slower release kinetics can be achieved if the mesoporous silica wall is functionalized with different functional groups. Among the existing organic groups, the functionalization with amino moieties has been widely reported.

The surface area has also been used to attach ligands that could target MSN to tumor (reviewed in [8]). For example, folate was attached on the surface of MSN to target the nanoparticles to cancer cells that have increased expression of folate receptor. Likewise, transferrin is attached to the MSN surface to target to cancer cells that have increased expression of transferrin receptor. Attachment of a cyclic peptide for integrin has also been used. Surface coating by hyaluronic acid (HA) is used to target cancer stem like cells. Together with passive targeting mechanism represented by the enhanced permeability retention (EPR) effect, MSNs can be targeted to tumor. The targeting ability of these new nanodevices should lead to tailor-made dosing regimens, with a significant reduction of severe side effects associated with some diseases such as cancer, and should also eventually result in a more efficient allocation of health care resources. Moreover, the use of these engineered products may allow the combination of the therapeutic potential and nanoscale diagnosis.

Surface coating by positively charged polymers such as PEI (polyetheleneimine) has been used to attract nucleic acids that are negatively charged [9]. In fact, the binding of nucleic acids is quite strong so that they can be protected from nuclease actions. This was used to deliver plasmid DNA to express a gene, and to deliver short interfering RNA (RNAi) to shut down gene expression. Lipid surface coating has also been pursued by a number of groups [10, 11]. In one example, the nanoparticles produced are called protocell, as the material resembles cells with lipid bilayer [10].

Instead of surface modification, additional feature can be incorporated into MSN nanoparticles. For example, magnetic core such as iron oxide has been encapsulated into inorganic silica mesoporous nanoparticles [12]. Iron oxide has superparamagnetic effect so that the exposure of such nanoparticles to oscillating magnetic field generates heat. This could be used for cancer applications, hence devising smart nanoparticles able to simultaneous and synergistically perform a dual function, to deliver cytotoxic agents in a controlled way and to give off heat (hyperthermia). In addition, MSNs containing iron oxide core can be used for enhancing MRI.

4. NANOMACHINES AND CONTROLLED DRUG RELEASE

Another key feature of MSNs is that they are relatively stable enabling the attachment of a variety of nanomachines resulting in the development of mechanized nanoparticles (Fig. 3). In this approach, the idea of nanovalve [13] or a cap [14] set the stage for dramatic development of various MSNs equipped with nanomachines, nanovalves are attached at the openings of pores to provide an open and close function. This enables controlled release of drugs that are stored inside the pores. Controlled release of anticancer drugs is important in cancer therapy, as this enables spatial and temporal control over drug release.

Various stimuli have been employed including internal conditions such as reducing condition, low pH and enzyme levels. In addition, mechanized nanoparticles that respond to external stimuli such as light, magnetic field and ultrasound have been developed (reviewed in [15, 16]).

A different approach is to modify pore walls so that movement can be generated inside the pore. An example is nanoimpeller that has azobenzene lining up the pore wall [17]. Since the conformation of azobenzene changes upon exposure to light, movement can be generated within the pore and this results in pushing the drug that is stored in the pore out of the nanoparticle.

5. BIOMEDICAL APPLICATIONS

Although mesoporous silica materials were initially designed for catalysis applications, it was soon recognized that their utility can be expanded to include other applications in diverse research areas, such as magnetism, sensors, optical materials, photo catalysis, fuel cells, thermoelectrics and even in health research.

In 2001, these materials were proposed as drug delivery systems for the first time [18]. The very same mesoporous ceramic matrices that were able to transform alcohol in gasoline were employed to adsorb and release pharmaceutical agents. Since then, a variety of drugs have been loaded onto MSNs and used for biomedical applications. Cancer application, infectious disease application as well as application for bone diseases are some of the major area of studies. In addition to therapeutic applications, MSNs were found to be effective as imaging agents. Various fluorescent dyes have been incorporated into MSNs. Also, MSNs have been used as imaging enhancing agents as described above briefly.

For biomedical application, the lack of toxicity is a prerequisite to employ these materials as drug delivery systems. Maximum tolerated dose of MSN was determined in mouse studies and short-term and long-term toxicity studies were carried out [19]. Excretion of MSN from animal body as well as the route of excretion has been investigated.

Equally important is the issue of tumor accumulation for cancer application. This type of study was carried out using xenograft mice (for example [19]). Fluorescent imaging, ICP-MS and the use of radioactively labeled MSNs were employed for such studies. The outcome appears to be determined by the property of MSNs used and detailed survey of literature is needed to achieve general conclusion.

The issue of material safety was of major driving force in the recent development of biodegradable PMO that defines a new generation of MSNs with favorable property for clinical translation (reviewed in [6]). While MSNs are degradable, it is desirable to tune this degradability. Thus, researchers have incorporated chemical bonds such as disulfide, tetrasulfide or peptide-like bonds into the framework of nanoparticles. This enables breakdown of the materials inside the body. The degraded materials are expected to be excreted out of the body. This is quite attractive, as the material used for biomedical application will disappear completely from the patient body.

The development of novel advanced multifunctional materials for a broad range of technological applications represents a renewed hope in many different fields. In particular, recent research breakthroughs in the biomedical arena have emerged as the basis for future personalized treatments and diagnostic techniques with a hitherto unsuspected selectivity. With the amazing advances in the preparation and characterization techniques of nanotechnology products, the possibility of manufacturing devices capable of establishing an intimate interaction with the biological world has been opened. This fact represents a precise control over the release processes of therapeutic substances and provides an opportunity to improve the specificity of the therapeutic action, as well as to reconsider some of the promising drugs for certain diseases that were once discarded by their low levels of tolerance.

6. OTHER SILICA-BASED MATERIALS

Silica materials that are non-porous have been used widely in biological studies. Among these, silica nanoparticles containing thiol groups have received interest, as various dyes can be attached to the thiol group and they can be used for imaging studies. Multilayered MSNs have been developed and they may provide important new materials in future applications.

ACKNOWLEDGMENTS

This work was supported by a Grant from JSPS KAKENHI Grant Number JP15K21764 (to F.T.) and a Grant from European Research Council (Advanced Grant VERDI: ERC-2015-AdG No. 694160) (to M.V.).

REFERENCES

[1] T. Yanagisawa, T. Shimizu, K. Kuroda, et al., The preparation of alkyltriethylammonium–kanenite complexes and their conversion to microporous materials, Bull. Chem. Soc. Jpn. 63 (1990) 988–992.
[2] C.T. Kresge, M.E. Leonowicz, W.J. Roth, et al., Ordered mesoporous molecular sieves synthesized by a liquid–crystal template mechanism, Nature 359 (1992) 710–712.
[3] Y. Sakamoto, S. Inagaki, T. Ohsuna, et al., Structure analysis of mesoporous material FSM-16—studies by electron microscopy and X-ray diffraction, Microporous Mesoporous Mater. 21 (1998) 589–596.
[4] D. Zhao, J. Feng, Q. Huo, et al., Triblock copolymer syntheses of mesoporous silica with periodic 50 to 300 angstrom pores, Science 279 (1998) 548–552.
[5] S. Inagaki, S. Guan, Y. Fukushima, et al., Novel mesoporous materials with a uniform distribution of organic groups and inorganic oxide in their frameworks, J. Am. Chem. Soc. 121 (1999) 9611–9614.
[6] J. Croissant, X. Cattoen, M. Wong Chi Man, Syntheses and applications of periodic mesoporous organosilica nanoparticles, Nanoscale 7 (2015) 20316–20334.
[7] J. Hu, M. Chen, X. Fang, et al., Fabrication and application of inorganic hollow spheres, Chem. Soc. Rev. 40 (2011) 5472–5491.
[8] R.E. Yanes, F. Tamanoi, Development of mesoporous silica nanomaterials as a vehicle for anti-cancer drug delivery, Ther. Deliv. 3 (2012) 389–404.
[9] C. Hom, J. Lu, M. Liong, et al., Mesoporous silica nanoparticles facilitate delivery of siRNA to shutdown signaling pathways in mammalian cells, Small 6 (2010) 1185–1190.
[10] J. Liu, A. Stace-Naughton, X. Jiang, et al., Porous nanoparticle supported lipid bilayers (protocells) as delivery vehicles, J. Am. Chem. Soc. 131 (2009) 1354–1355.

[11] H. Meng, M. Wang, H. Liu, et al., Use of a lipid-coated mesoporous silica nanoparticle platform for synergistic gemcitabine and paclitaxel delivery to human pancreatic cancer in mice, ACS Nano 9 (2015) 3540–3557.
[12] M. Liong, J. Lu, M. Kovochich, et al., Multifunctional inorganic nanoparticles for imaging, targeting and drug delivery, ACS Nano 2 (2008) 889–896.
[13] R. Hernandez, H.R. Tseng, J.W. Wong, et al., An operational supramolecular nanovalve, J. Am. Chem. Soc. 126 (2004) 3370–3371.
[14] C. Lai, B. Trewyn, D. Jefinija, et al., A mesoporous silica nanosphere-based carrier system with chemically removable CdS nanoparticle caps for stimuli-responsive controlled release of neurotransmitters and drug molecules, J. Am. Chem. Soc. 125 (2003) 4451–4459.
[15] H. Mekaru, J. Lu, F. Tamanoi, Development of mesoporous silica-based nanoparticles with controlled release capability for cancer therapy, Adv. Drug Deliv. Rev. 95 (2015) 40–49.
[16] A. Baeza, M. Manzano, M. Colilla, et al., Recent advances in mesoporous silica nanoparticles for antitumor therapy: our contribution, Biomater. Sci. 4 (2016) 803–813.
[17] J. Lu, E. Choi, F. Tamanoi, et al., Light-activated nanoimpeller-controlled drug release in cancer cells, Small 4 (2008) 421–426.
[18] M. Vallet-Regi, A. Ramila, R.P. del Real, et al., A new property of MSM-41: drug delivery system, Chem. Mater. 13 (2001) 308–311.
[19] J. Lu, M. Liong, Z. Li, et al., Biocompatibility, biodistribution and drug-delivery efficacy of mesoporous silica nanoparticles for cancer therapy in animals, Small 6 (2010) 1794–1805.

CHAPTER TWO

Structure Characterization of Mesoporous Materials by Electron Microscopy

Yanhang Ma*, Yaping Zhang*, Lu Han†, Shunai Che†,‡, Osamu Terasaki*,1

*School of Physical Science and Technology, ShanghaiTech University, Pudong, Shanghai, China
†School of Chemical Science and Engineering, Tongji University, Shanghai, China
‡School of Chemistry and Chemical Engineering, State Key Laboratory of Metal Matrix Composites, Shanghai Jiao Tong University, Shanghai, China
[1]Corresponding author: e-mail address: osamuterasaki@mac.com

Contents

Abstract

The structural analysis of silica mesoporous crystals (SMCs) is a key step for understanding their properties, especially toward drug delivery. Although various techniques can be applied, electron crystallography has been proved the most powerful method in the structure solution of mesoporous materials. In this chapter, we will begin with introducing basic crystallography and extend to the 3D structure solution using a set of HRTEM images. Typical examples will also be given to show the structure features of 2D, 3D, cage-type, and hyperbolic surface-type mesostructures. A drug delivery system using SMCs as a reservoir will be given at the end.

1. INTRODUCTION

If a material shows a sharp diffraction pattern, then it is a crystal following IUCr definition. Here, for simplicity we define a crystal as a repeating moiety (pattern). Nanoporous crystals are classified into three types

The Enzymes, Volume 43
ISSN 1874-6047
https://doi.org/10.1016/bs.enz.2018.07.002

depending on their pore diameter d, microporous: $d < 20$ Å, mesoporous: 20 Å $\leq d \leq$ 500 Å, and macroporous: $d >$ 500 Å, respectively. Here we confine our discussion within silica mesoporous crystals (SMCs). SMCs have a periodic arrangement of pore/channel which are supported with amorphous silica walls in contrast to conventional inorganic crystals, however SMCs show characteristic crystal morphology, which is commensurate with their point group symmetries. In order to meet the target of this book, we will try to describe characteristic features of SMCs' structure. Fine structures of surface, defects of crystals, and a method how to observe opening mouth of the pores on the external surfaces are very important for using SMCs as drug carriers, please read the following reference for recent advancements (Springer Handbook of Microscopy, to be published). Here we will keep the procedure for solving the structure in minimum level.

SMCs are the most typical mesoporous crystal and are normally formed through self-assembly of amphiphilic molecules (surfactants or block copolymer) in water as a template (structure directing agent) for subsequent and/or simultaneous condensation of water soluble inorganic silica source at the boundary of water and the amphiphilic molecules. We can see therefore similarities between SMCs and the lyotropic liquid crystals (LLCs). LLCs are formed by amphiphilic molecules into bilayer, cylindrical, and spherical phases in the presence of water. The structure of LLCs is generally described by either surfactant packing parameter g ($g = V/a_0 l$, where V is the surfactant chain volume, a_0 is the effective hydrophobic/hydrophilic interfacial area, and l is chain length) or curvature of the interface between water and amphiphilic molecules. While, the synthesis-field diagram for SMCs is more complex and diverse compared to that for the LLCs due to the silica condensation. Structures of SMCs change from cage-type with different shapes, 2D-cylindrical, bicontinuous cubic or tricontinuous hexagonal, and 1D-lamellar with increase of the packing parameter from 1/3 to 1.

2. BASIC PROCESS FOR OBTAINING STRUCTURE OF SMCs

Periodic lattice $L(\boldsymbol{r})$ is given by an infinite array of points in a real space using Dirac delta function,

$$L(\boldsymbol{r}) = \sum_{\boldsymbol{r}_n} \delta(\boldsymbol{r} - \boldsymbol{r}_n) \tag{1}$$

where any lattice point $\boldsymbol{r}_n$ has a center of symmetry and is given by a linear combination of primitive translational (basis) vectors, $\boldsymbol{a}_\mathrm{i}$,

$$r_n = \sum_{i=1}^{m} n_i \boldsymbol{a}_i \tag{2}$$

where $n_i (1 \leq i \leq m,\ m \leq 3)$ are arbitrary integers.

Any position $\boldsymbol{r}$ in the crystal is given by the origin, $\boldsymbol{r}_m$, of the mth unit cell and its coordinates (x, y, z) within the cell as:

$$\begin{aligned}
&\text{For 1D case}, n=1,\ \boldsymbol{r}=\boldsymbol{r}_m + x\boldsymbol{a}_1 \\
&\text{For 2D case}, n=2,\ \boldsymbol{r}=\boldsymbol{r}_m + x\boldsymbol{a}_1 + y\boldsymbol{a}_2 \\
&\text{For 3D case}, n=3,\ \boldsymbol{r}=\boldsymbol{r}_m + x\boldsymbol{a}_1 + y\boldsymbol{a}_2 + z\boldsymbol{a}_3
\end{aligned} \tag{3}$$

where $0 \leq x, y, z \leq 1$.

Each reciprocal lattice point, hkl, in reciprocal space corresponds to a set of lattice planes in real space and is given by the reciprocal lattice vectors, $\boldsymbol{b}_i$,

$$\boldsymbol{K}_{hkl} = h\boldsymbol{b}_1 + k\boldsymbol{b}_2 + l\boldsymbol{b}_3 \tag{4}$$

The inter-planar spacing d_{hkl} is given by

$$d_{hkl} = 2\pi / |\boldsymbol{K}_{hkl}| \tag{5}$$

A crystal structure $C(\boldsymbol{r})$ is given by subtracting a basis $B(\boldsymbol{r})$ convoluted with the periodic lattice $L(\boldsymbol{r})$ of infinite size from the infinite space $A(\boldsymbol{r})$.

$$C(\boldsymbol{r}) = A(\boldsymbol{r}) - B(\boldsymbol{r}) * L(\boldsymbol{r}) \tag{6}$$

For a finite crystal, a shape function, given by $Z(\boldsymbol{r}) = 1$ when $\boldsymbol{r}$ is inside the crystal and $Z(\boldsymbol{r}) = 0$ otherwise, is introduced to handle any size and shape. Then the crystal $C(\boldsymbol{r})$ should be defined by the following general equation:

$$C(\boldsymbol{r}) = \left[A(r) - B(\boldsymbol{r}) * L(\boldsymbol{r})\right] \cdot Z(\boldsymbol{r}) \tag{7}$$

and a crystal structure factor (CSF) $F(\boldsymbol{K}_{hkl})$ is defined with Fourier transform of a crystal as:

$$F(\boldsymbol{K}_{hkl}) = \mathcal{F}[C(\boldsymbol{r})] = \{\mathcal{F}[A(\boldsymbol{r})] - \mathcal{F}[B(\boldsymbol{r})] \cdot \mathcal{F}[L(\boldsymbol{r})]\} * \mathcal{F}[Z(\boldsymbol{r})] \tag{8}$$

and $\mathcal{F}[Z(\boldsymbol{r})]$ is Fourier transform of crystal shape (geometrical) function [1].

It is clear from Eq. (8) that once we have obtained a set of crystal structure factors $F(\boldsymbol{K}_{hkl})$ for various hkl reflections, $\boldsymbol{K}_{hkl}$, through diffraction measurement or Fourier transform of HRTEM images, a crystal structure $C(\boldsymbol{r})$ can be obtained through inverse Fourier transformation.

3. 2d-SMCs

It is not so easy to solve a structure of SMCs solely from powder XRD data even for the relatively simple 2d-structure. This is because $\mathcal{F}[B(\boldsymbol{r})]$ dumps rapidly as the magnitude of $\boldsymbol{K}_{hkl}$ increases. However, the structural features of 2d-SMCs can be easily observed by TEM images, if they are taken along the channel direction except for chiral case. The HRTEM image of FSM-16 shows highly ordered mesopores with six-fold symmetry along the channel direction, which is difficult to reconcile with the originally proposed "folded sheet" formation mechanism. However, another type of 2S-SMC KSW-2 with *c2mm* symmetry indicates "folded sheet" mechanism is operating as shown in Fig. 1 [2, 3]. This part may be fully discussed in another chapter "Mesoporous silica nanomaterials synthesis" by Kuroda in The Enzymes, Volume 44.

Simple 2d-SMC is either MCM-41 or SBA-15. Both structures are well described by the pores (rather than pipes) arranged on a hexagonal

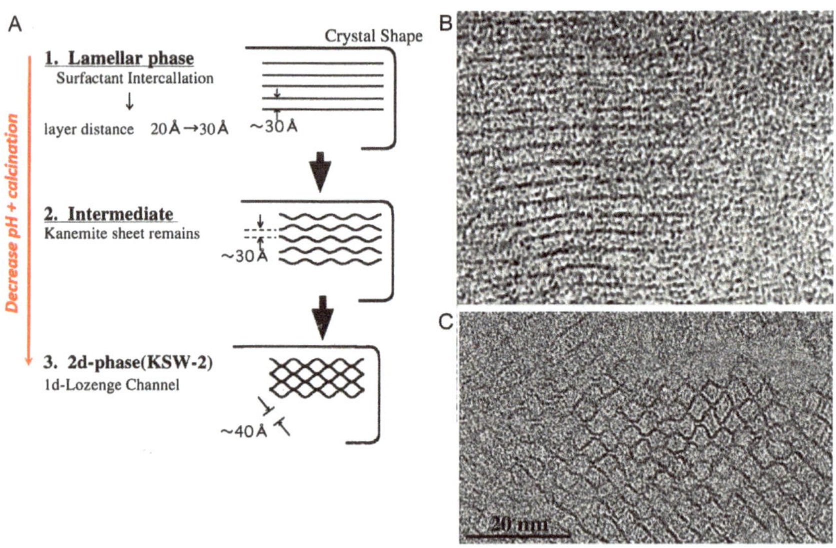

Fig. 1 (A) Schematic drawing of kanemite sheet and the intercalating effect with surfactant. (B) HRTEM image of lamellar phase and (C) HRTEM image of KSW-2 with lozenge shaped channel. Structure change from lamellar phase (B) to lozenge shape (C) through reducing number of surfactant between the layers is schematically drawn in the left column.

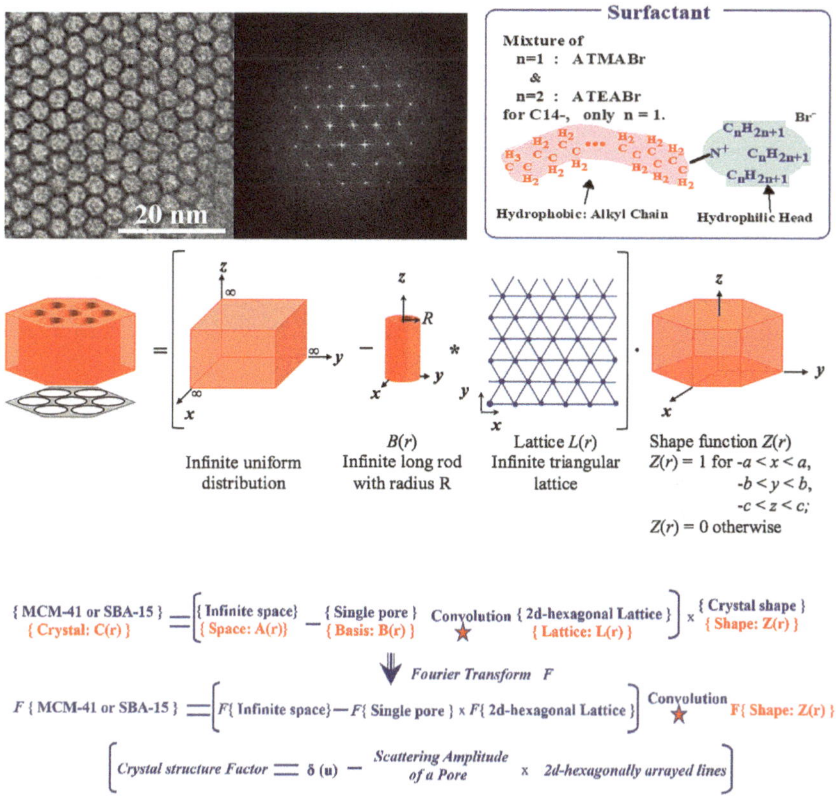

Fig. 2 Corresponding to Eq. (7), crystal of MCM-41 or SBA-15 can be described as single pore ($B(\mathbf{r})$) is convoluted with 2d-hexagonal lattice ($L(\mathbf{r})$). The highest point group for 2d-crystal system is *p6mm*.

lattice (Fig. 2). Supporter of pores is continuous silica wall of SMCs (flesh-color in Fig. 2), however, silica pore-surface is not smooth even in MCM-41. In the case of SBA-15, we can (i) control roughness of the surface and (ii) furthermore introduce holes in the silica wall to connect neighboring pores. These have been studied by "in situ" gas adsorption XRD experiments [4].

Chiral 2d-SMC is a very interesting material, however it requires more elaborated discussion which is beyond of this scope. 2d-Periodic mesoporous organosilicas [5] is another exciting SMC and this will be discussed in the chapter "Periodic mesoporous organosilica" by Inagaki in The Enzymes, Volume 44.

4. 3d-SMCs

For 3d-case, if one will synthesize a new SMC, following two experimental steps are strongly recommended simultaneously:

(i) to observe powder X-ray diffraction (XRD) pattern and

(ii) to observe a SEM image for checking overall sample whether a sample looks uniform from appearance and for checking existence of any particles with facets.

Then one may move to observe electron diffraction (ED) pattern and/or TEM image of a particle with clear facets.

Importance of the above processes can be shown with an example of SMCs called AMS-9. If you observe this powder XRD pattern only (Fig. 3A), you might give up for further study. However, the crystal shows a nice morphology with facets indicating highly crystalline state (SEM image, Fig. 3B), furthermore the HRTEM image shows beautiful repetitions of structure units (Fig. 3C). It is clear from the HRTEM image that this AMS-9 is a beautiful crystal with a very large unit cell marked by a red square. The FD obtained from the HRTEM image shows very well-defined array of spots with fourfold symmetry (Fig. 3D). Fig. 3B and D suggests this crystal belongs to a tetragonal system. The rotational integrated pattern of *hk*0 components (Fig. 3D) among general *hkl* reflections gives Debye–Scherrer type ring-pattern and its 1d-intensity profile through origin shows well-separated diffraction profile (Fig. 3E). Fig. 3E should be compared with powder XRD pattern (Fig. 3A) although Fig. 3A contains not only *hk*0 but also all *hkl* reflections. A set of these data clearly show that FD/ED pattern can give much higher resolution than powder XRD pattern. Especially, products of a novel SMC at early stage may contain more than one phase and size of novel one might be very small, then electron microscopy will play a unique and powerful role.

Fourier transform of the HRTEM image called FD (Fig. 3D) gives a set of Fourier components of the crystal $C(\boldsymbol{r})$ called CSFs in complex number, $F(\boldsymbol{K}_{hkl}) = Abs[F(\boldsymbol{K}_{hkl})]e^{i\alpha(hkl)}$. Once the origin is chosen at the right place in the images taken with different incidences, all phases $\alpha(hkl)$ are uniquely determined. Fig. 3D shows only amplitude terms of CSFs, $Abs[F(\boldsymbol{K}hkl)]$, but phase term $\alpha(hkl)$ stored in computer memories will play an essentially important role for constructing the 3D-electrostatic potential distribution map. In order to reduce multiple scattering effect, we should obtain FDs from the HRTEM images taken at very thin region. Then the 3D-electrostatic

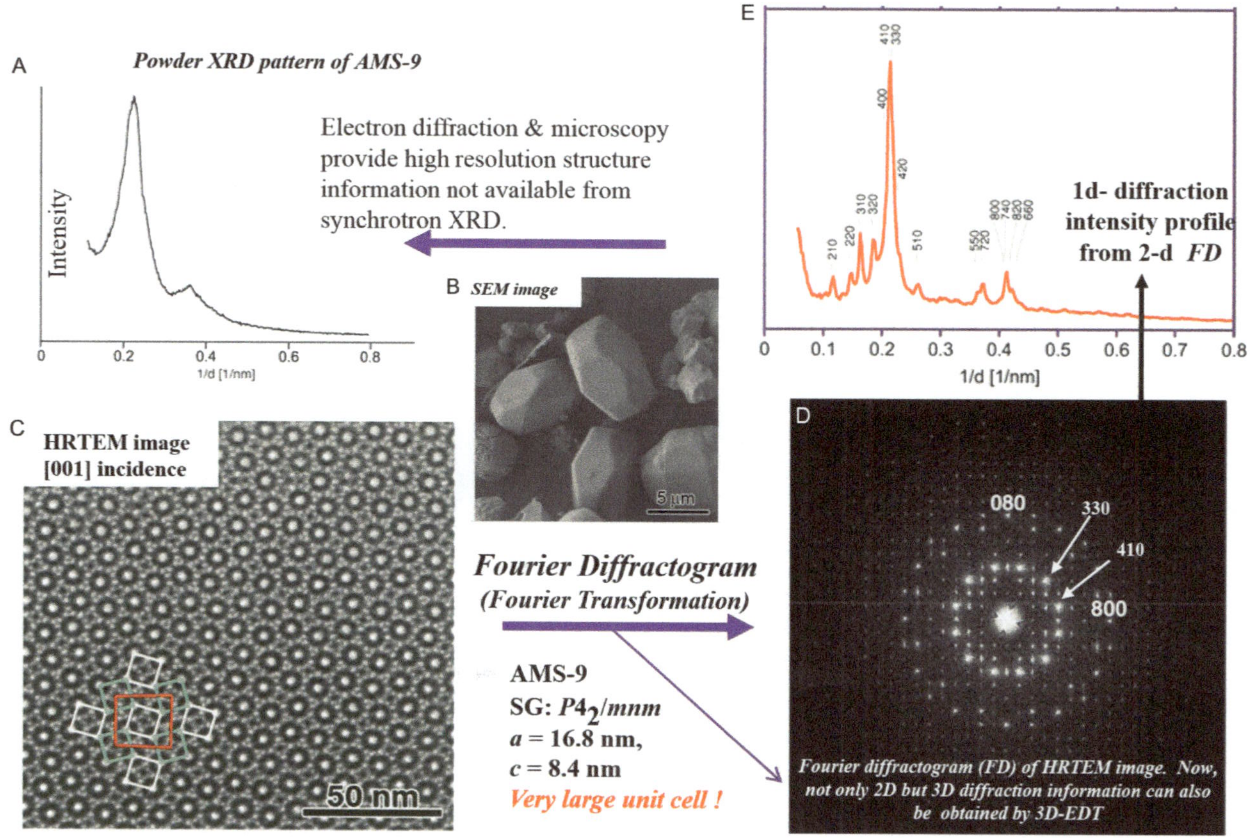

Fig. 3 Power of electron microscopy is shown using AMS-9 as an example for structure determination of novel SMCs. Powder XRD pattern (A), SEM image (B), HRTEM image along [001] (C), FD taken from (C) and (D), rotational integrated intensity of (D) and (E).

potential distribution map can be obtained from Fourier sum of CSFs and the map gives crystal structure of the SMCs uniquely, although resolution is limited by maximum *q*-value $\left(q = \frac{4\pi sin\theta}{\lambda}\right)$ of reflections given by the resolution of the images. In contrast, structure solution through traditional crystallography based on diffraction intensity measurement such as X-ray or Neutron can give higher resolution but the solution is the most probable one imposed by uncertainty of phase information of CSFs. The flowchart for structure solution of SMCs is given in Fig. 4.

Let's move to show a summary of the structures of SMCs, most of which were solved by us after developing methodology, as a set of HRTEM images, FD patterns, and structure drawings (Fig. 5). These structures can be categorized into three types: cage-type (both spherical and polyhedral packing), cylindrical (2d-SMCs), and 3d-periodic minimal surface type as the increase of interfacial curvature. 3d-Hexagonal $P6_3/mmc$ has a feature of both spherical and polyhedral packing. Please enjoy beautiful HRTEM images.

Up to now, 2d-SMC SBA-15 has been widely used for drug carrier, however, I believe SMCs of 3d-cage type or of hyperbolic surface type will be more valuable than 2d-SMCs for carrying a large amount of drugs and keeping a pathway of drug release opening from clogging of 1d-pores (Table 1).

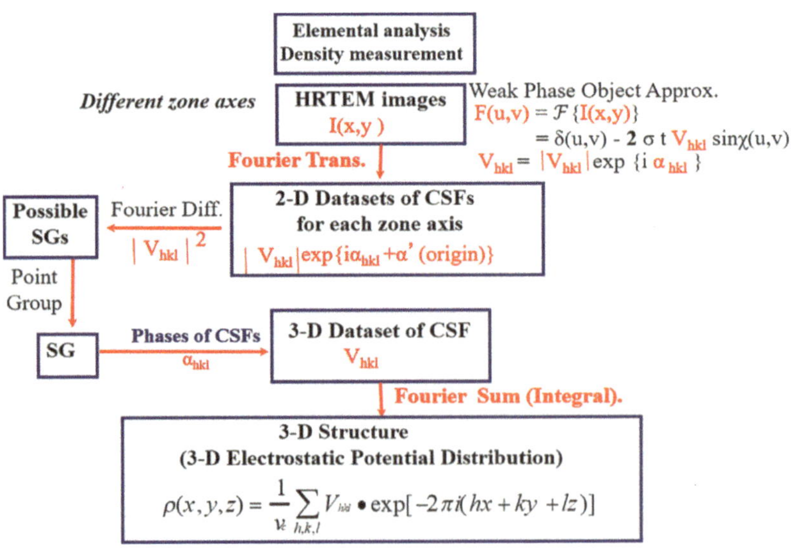

Fig. 4 3D structure solution from a set of HRTEM images.

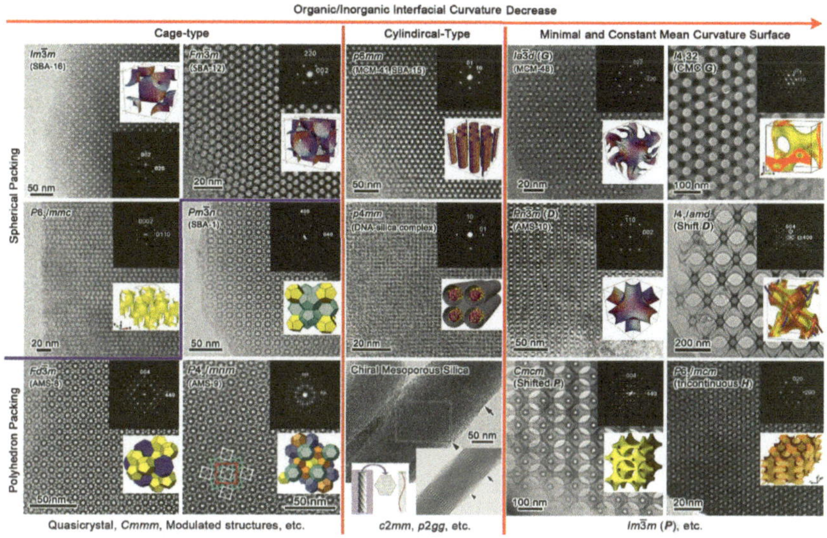

Fig. 5 Summary of the structures of SMCs.

In order to synthesize SMCs, it is important to complete synthesis-field diagram as shown in Fig. 6. SMCs using anionic surfactant called AMS-*n* ($n = 1$–10) have been extensively developed by Shunai Che and her group, Shanghai Jiao Tong University [23]. Many structure types have been obtained by changing experimental parameters (Fig. 6). Furthermore, SEM images and TEM images obtained from AMS-*n* are beautiful even though some of AMS-*n* structures are synthesized in kinetic control.

We will discuss important crystallographic aspects of cage-type and minimal surface type of SMCs SBA-16 (*Im*-3*m*) and MCM-48 (*Ia*-3*d*) in details.

5. CAGE-TYPE SMCs

The cage-type SMCs are formed by the packing of the spherical or polyhedral micelles with the highest organic/inorganic interfacial curvature. The cage-type mesostructures is analog to the traditional atomistic crystal formed by the packing of atoms. Up to now, different cage-type mesostructures have been fabricated, providing us the opportunity to look into the structural relationships of their packing behaviors. Especially, compared to the traditional atomistic crystals, the intergrowth and defects of SMCs are more diverse due to the flexibility of the silica wall before completion of silica condensation so that the structures are sensitive to the

Table 1 Structure Details Shown in Fig. 5 Are Summarized Here

Series	Type	Material	Description	Space Group	Lattice Const (Å)	Pore/Channel Size (Å)	Aperture Size (Å)
1-1 [6]	Cage	SBA-16		$Im\bar{3}m$	133	95	23 [111]
1-2 [7]		CFMS[a]	a[b]	$Fm\bar{3}m$	90.7	32.3	
			b	$Fm\bar{3}m$	94.5	32.3	
2-1 [8]		SMC	As-made	$P6_3/mmc$	$a = 60.6$ $c = 98.1$		
			Calcined		$a = 55.8$ $c = 91.2$	42	
1-3 [9, 10]	Cylindrical	MCM-41 [9]		$p6mm$	45	30	
		SBA-15 [10]	As-synthesized	$p6mm$	120		
			Calcined	$p6mm$	86.0–136	46–100	
2-3 [11]		DSC	DNA-silica complex	$p4mm$	23		
3-3 [12]		Chiral mesoporous silica	Chiral	$p6mm$	44	22	

2-2 [6]	Polyhedron packing	SBA-1	As-synthesized	$Pm\bar{3}n$	86		
			Calcined		73	HREM: A: 15 × 22, B: 2 BET: A:24, B:8	
	Polyhedron packing	SBA-6	As-synthesized	$Pm\bar{3}n$	160		
			Calcined		146	A:85 B:73	A-B: 20 A-A: 33 × 41
3-1 [13]		AMS-8	From TEM	$Fd\bar{3}m$	177.9	L: 75–76 S: 52–56	L-L: 14–25 L-S: <5 S-S: <5
3-2 [14]	Polyhedron packing	AMS-9		$P4_2/mnm$	$a = 84.0$ $c = 167.9$	38.3	
1-4 [15]	Mean curvature surface	SMC (G-type)		$Ia\bar{3}d$	136	42.9	
1-5 [16]		SG	Single gyroid	$I4_132$	~700		
2-4 [17]		AMS-10		$Pn\bar{3}m$	96	47	
2-5 [18]		MS-1	Mass (THF/H_2O) = 4	$I4_1/amd$	$a = 1680$ $c = 2376$	~1000	100–150
		MS-2	Mass (THF/H_2O) = 8		$a = 2300–2500$ $c = 3250–3535$	~1400	380–450
3-4 [19]		SDP	Framework shift	$Cmcm$	$a = b = 1480$ $c = 1050$		
3-5 [20]		IBN-9		$P6_3/mcm$	$a = 88.4$ $c = 84.3$	30–32	

[a]CFMS: Carboxylic group functionalized mesoporous silica.
[b]a, b use $[C_{18}H_{37}N(CH_3)_2(CH_2)_3N(CH_3)_3]\ Br_2(C_{18\text{-}3\text{-}1})$ as the template and use the Sur:CES:HCl:TEOS:H_2O different ratios.

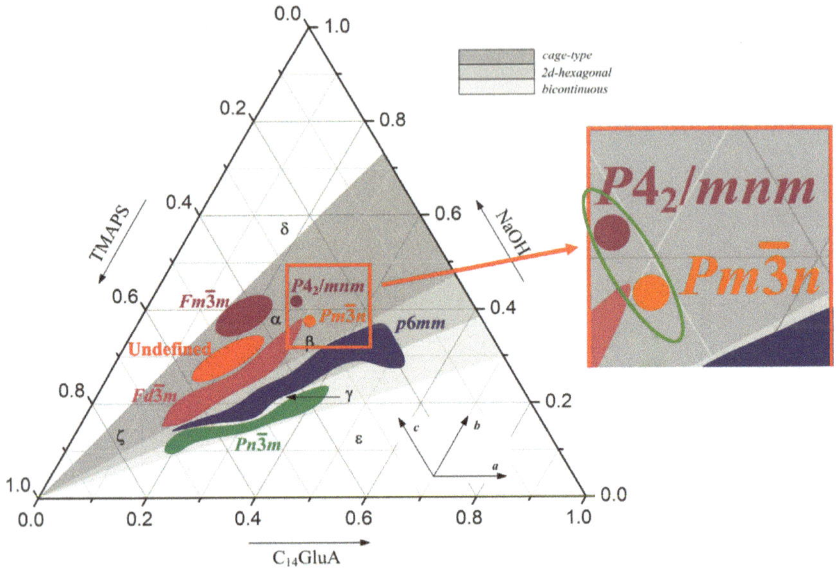

Fig. 6 Synthesis-field diagram (mole fraction) of the C_{14}GluA–NaOH–TMAPS system [21]. Dodecagonal quasicrystal was successfully synthesized in the marked area near $P4_2/mnm$ (AMs-9) and Pm-3n (AMS-8) marked with green ellipse [22].

synthesis conditions. Up to now, several cage-type structures have been solved by electron microscopy, for instance, cubic *Im*-3*m* (SBA-16, FDU-1), cubic *Fm*-3*m* (SBA-12, KIT-5, FDU-12), cubic *Pm*-3*n* (SBA-1, SBA-6), cubic *Fd*-3*m* (FDU-2, AMS-8), hexagonal $P6_3/mmc$ (SBA-2, SBA-11), orthorhombic *Pmmm* (FDU-13), tetragonal *P*4/*mmm* (FDU-11), tetragonal $P4_2/mnm$ (AMS-9), etc.

Structure of SBA-16 is the simplest cage-type SMC and very valuable to understand simple crystallography. It shows clear morphology (Fig. 7) and gives beautiful HRTEM images (Fig. 8). SBA-16 is formed in the following: one type of sphere with diameter $d, d = a\sqrt{3}/2$, is arranged in body-centered cubic (*bcc*) lattice with space group of *Im*-3*m* and unit cell constant *a*. The point group symmetry is the highest *m*-3*m* (full symbol: $4/m\ -3\ 2/m$ which means fourfold rotation symmetry along $\langle 100 \rangle$ and mirror perpendicular to the axis, threefold rotoinversion symmetry along $\langle 111 \rangle$, and twofold rotation symmetry along $\langle 110 \rangle$, and mirror perpendicular to the axis).

There are two typical crystal morphologies, known as equilibrium and growth forms. In the case of equilibrium form, the crystal and its surroundings are in thermodynamic equilibrium, and morphology is governed by minimum surface energy under constant volume and is well described by

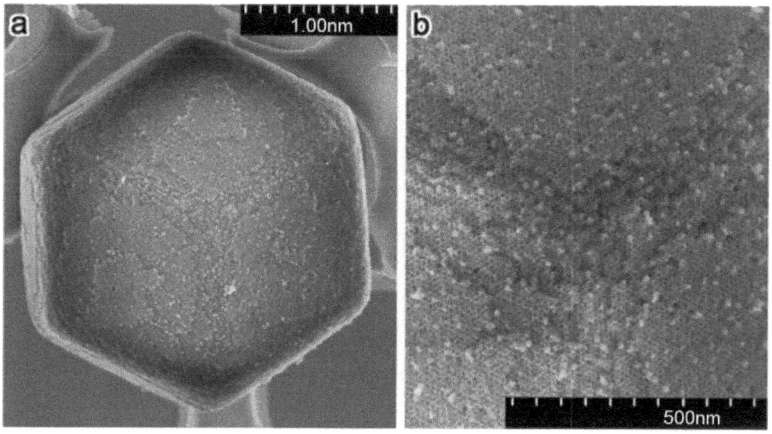

Fig. 7 SEM image of SBA-16, low magnification (A) and high resolution (B).

Wulff construction. The growth form is governed by anisotropy of growth rates in different crystallographic directions and strongly depends on growth conditions. In both cases, however, crystal morphologies should be commensurate with the point group symmetry of the crystal. Typical crystal morphologies of *m*-3*m* are cubic surrounded by {100} planes, octahedron by {111} planes, and dodecahedron by {110} planes, truncated octahedron by {100} and {111} planes and so on.

Using low-impact energy electrons, we are able to observe SEM images from electrically insulating materials without any metallic coatings. On overcoming, the resolution problem at low electron energy, surface fine structures include not only growth steps but also channel and cage openings, as well as their arrangement on the external surface have been observed [24].

For example, the channel openings, growth steps, and heights are observed in high resolution scanning electron microscopy (HRSEM) images of SBA-16 taken almost along the threefold axis, [111]. Fig. 7B is an enlarged image of Fig. 7A, allowing us to clearly see layer-by-layer growth steps which are common in various zeolites and SMCs (e.g., SBA-16), micropores in the silica wall of SBA-15 and so on [25].

Our method was developed especially for periodic structural arrangements with mesoscale by use of HRTEM images (Fig. 8A–C). As discussed earlier, HRTEM images give both the amplitudes and phases of the necessary CSFs. The 3d-space electrostatic potential map can be obtained as a unique structure solution through the inverse Fourier transformation (Fig. 8D and E). The boundary between silica wall and pores, which is

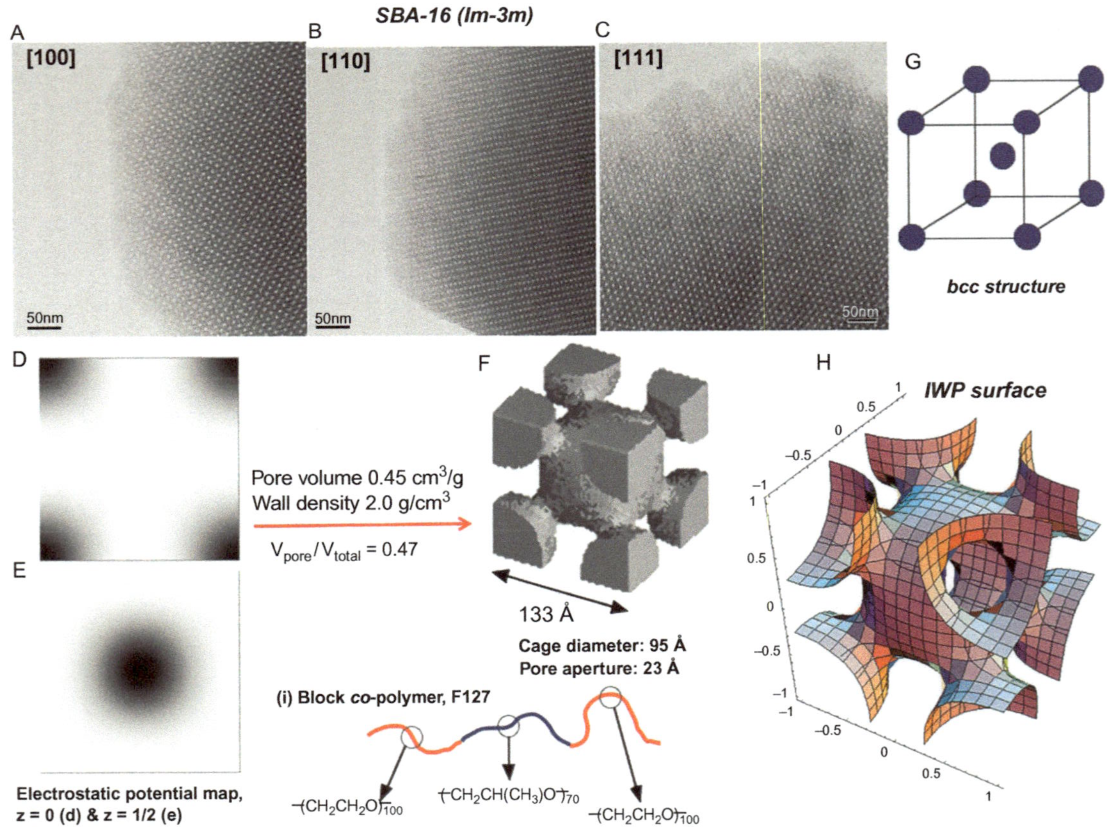

Fig. 8 High resolution TEM images of SAB-16 taken along [100], [110], and [111], (A), (B), and (C), respectively. Electrostatic map obtained from CSFs by Fourier inverse at $z=0$ (D) and $=1/2$ (E). 3D structure obtained from (D) and (E) combined with N_2 gas adsorption isotherm (F) and corresponding IWP-surface (H). A part of block *co*-polymer used for synthesis of SBA-16 (I).

constant-potential surface on the 3d-space electrostatic potential map, was determined through pore volume obtained from gas adsorption isotherm measurement and silica wall density. The structure of SBA-16 was determined (Fig. 8F).

In the case of SBA-16, surface of spherical cages is not smooth like SBA-15 and roughness of the surface was directly estimated by "in situ" gas adsorption XRD experiments [26]. There are many interesting crystallographic topics on defects in the cage-type SMCs, however they lie in different topics from this review.

6. HYPERBOLIC SURFACE TYPE

Next examples are hyperbolic surfaces SMCs. MCM-48 and AMS-10 [16] are typical ones with the space group of *Ia*-3*d* and *Pm*-3*n*, respectively. Their silica wall structures are well described with *G* (*Gyroid*)-surface and *D* (*Diamond*)-surface. Here we will discuss MCM-48 in details (Fig. 9).

Powder XRD pattern, SEM image, and one of HRTEM images, silica wall and two channel models are shown in Fig. 10. Truncated octahedron observed in SEM image indicates point symmetry *m*-3*m* (Fig. 10C). Extinction conditions, *hkl*: $h + k + l = 2n$, 0*kl*: $k, l = 2n$, *hhl*: $2h + l = 4n$, *h*00: $h = 4n$, according to Fourier diffractograms (FDs) obtained from a set of HRTEM images (Fig. 10C) uniquely suggests the space group *Ia*-3*d*. By the same procedure discussed for structure solution of SBA-16, the structure of MCM-48 was determined to be *G*-surface (Fig. 10D, left) and the *G*-surface separates left- and right-handed channels (Fig. 10D, right).

Complementary pores through the silica wall at the special flat point of the *G*-surface (SG: *Ia*-3*d*) on the threefold axis (16a site, site symmetry −3) was found in KIT-6 synthesized by Ryoo et al. (Fig. 11) [27].

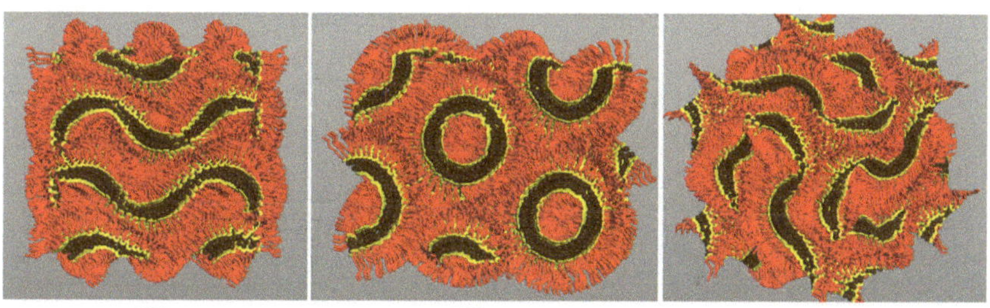

Fig. 9 Schematic drawing of surfactant molecules and silica wall for MCM-48. Hydrophilic head and hydrophobic tail groups of surfactant molecules and silica wall are shown by yellow, red, and black, respectively.

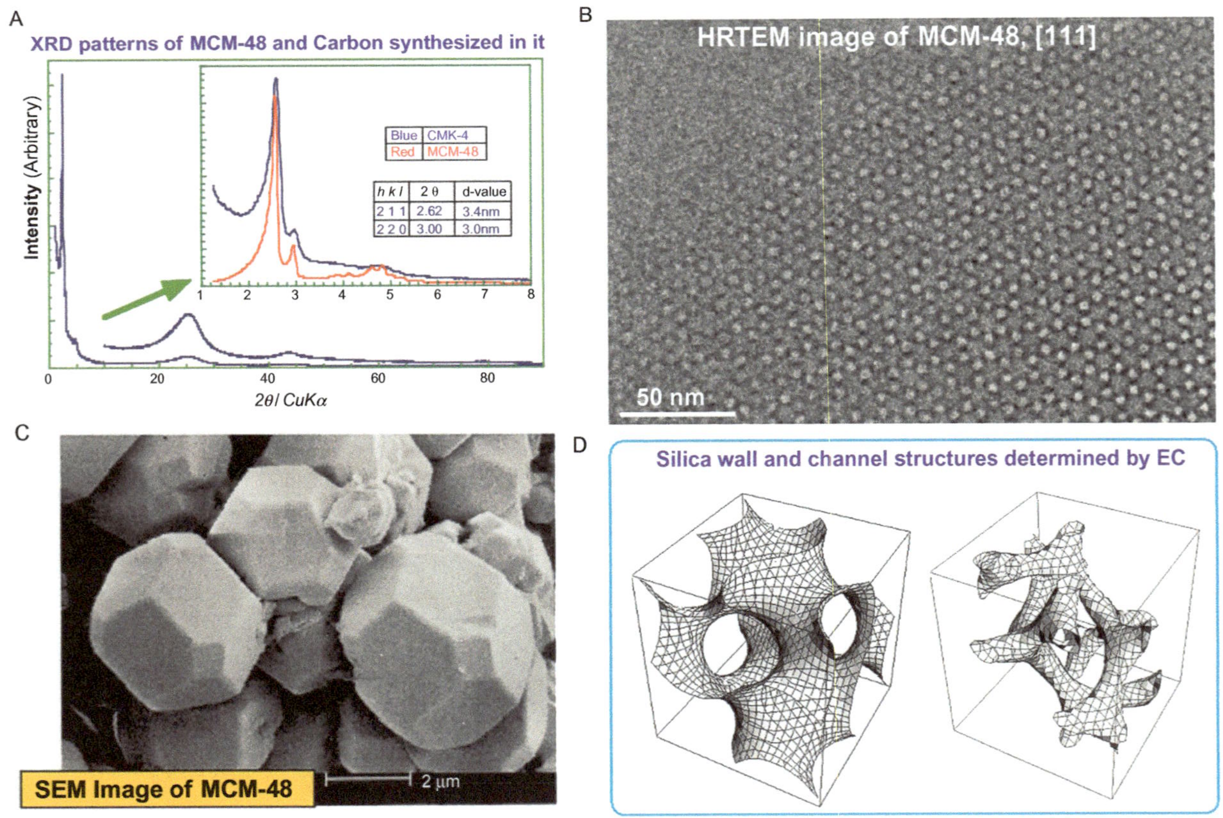

Fig. 10 PXRD pattern, SEM image, HRTEM image along [111] of MCM-48, and schematic representation of two channel models.

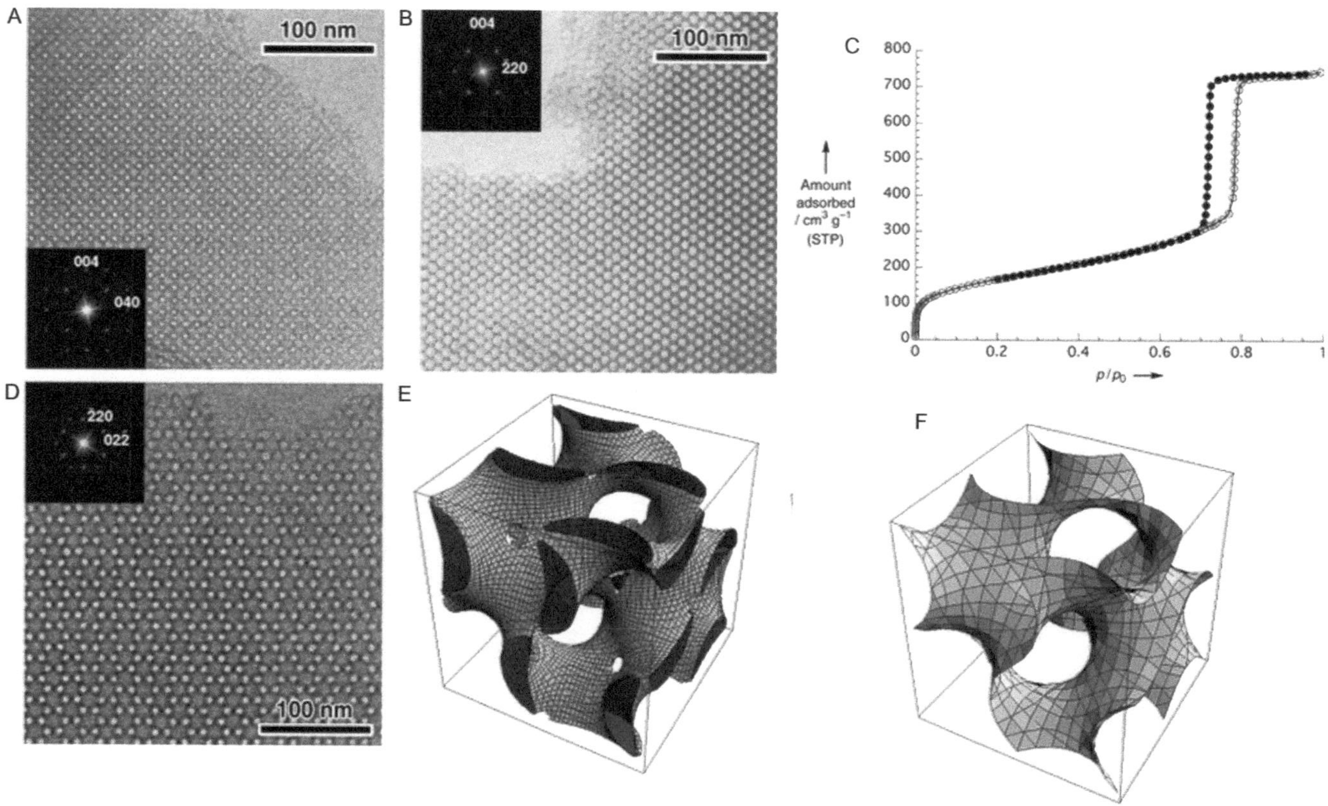

Fig. 11 HRTEM images of KIT-6 along [100] (A), [1−10] (B), [111] (D), N_2 adsorption/desorption isotherm (C), 3D distribution of silica wall and pores (E), and boundary between wall and pore (F).

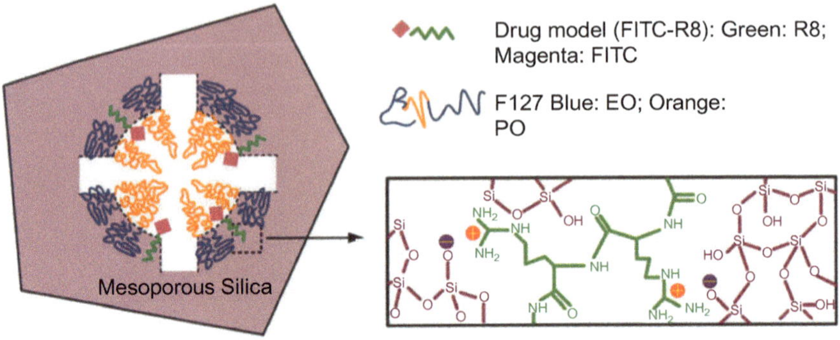

Fig. 12 Schematic drawing of the drug delivery system based on SMC loaded with FITC-R8.

Although a set of HRTEM images (Fig. 11A, B, and D) are very similar to those of MCM-48 (G-surface), we can confirm an existence of complementary pores with precise analysis. The size of the complementary pore is sensitive to the threshold value, which defines the boundary between the silica wall and the spaces in electrostatic potential distribution through N_2 gas adsorption isotherm data, and also dependent on silica wall density. The diameter is determined to be 17 Å by assuming silica wall density is 2.20 g/cm^3. The complementary pores produce connections between the left- and right-handed channels, then pore-system becomes interconnecting.

Due to the suitable pore size, high pore volume, and specific surface area, variable mesostructures and functional groups, SMCs become good candidates for drug storage and delivery. In fact, the use of SMCs in drug delivery has been widely studied. Notably, a controllable delivery-release system has been developed using SMC as a drug reservoir and peptide octarginine (R8) as a vector to endow the drug (fluorescein isothiocyanate, FITC) with cell-penetrating property. The SMC was synthesized using Pluronic F127 as template and FITC-R8 was incorporated by a one-pot synthesis route. The stable storage and sustainable release of drugs have been achieved due to the hydrophobic part of template molecule filling in the mesopores and the electrostatic interaction between the drug and siloxide ions that act as the diffuse-limiting factors for the drug diffusion [28] (Fig. 12).

ACKNOWLEDGMENTS

This work is supported by Recruitment Program for Foreign Experts, China (O.T.) and ShanghaiTech University (Y.M., Y.Z., and O.T.), the Commission for Science and Technology of Shanghai Municipality (17ZR1418600) and the Young Elite Scientist

Sponsorship Program by CAST (2017QNRC001) (Y.M.), National Natural Science Foundation of China (21571128), the National Excellent Doctoral Dissertation of PR China (201454), and the Shanghai Rising-Star Program (17QA1401700) (L.H.).

The authors acknowledge Sam M. Stevens, Exeter University, UK for English revisions.

REFERENCES

[1] L. Han, K. Miyasaka, O. Terasaki, Electron crystallography, in: D.W. Bruce, D. O'Hare, R.I. Walton (Eds.), Structure From Diffraction Methods, John Wiley & Sons Ltd, Chichester, 2014, pp. 201–258.
[2] Y. Sakamoto, S. Inagaki, T. Ohsuna, N. Ohnishi, Y. Fukushima, Y. Nozue, O. Terasaki, Structure analysis of mesoporous material 'FSM-16'—studies by electron microscopy and X-ray diffraction, Microporous Mesoporous Mater. 21 (1998) 589–596.
[3] T. Kimura, T. Kamata, M. Fuziwara, Y. Takano, M. Kaneda, Y. Sakamoto, O. Terasaki, Y. Sugahara, K. Kuroda, Formation of novel ordered mesoporous silicas with square channels and their direct observation by transmission electron microscopy, Angew. Chem. Int. Ed. 39 (2000) 3855–3859.
[4] N. Muroyama, A. Yoshimura, Y. Kubota, K. Miyasaka, T. Ohsuna, R. Ryoo, P.I. Ravikovitch, A.V. Neimark, M. Takata, O. Terasaki, Argon adsorption on MCM-41 mesoporous crystal studied by in situ synchrotron powder X-ray diffraction, J. Phys. Chem. C 112 (2008) 10803–10813.
[5] S. Inagaki, S. Guan, T. Ohsuna, O. Terasaki, An ordered mesoporous organosilica hybrid material with a crystal-like wall structure, Nature 416 (2002) 304–307.
[6] Y. Sakamoto, M. Kaneda, O. Terasaki, D.Y. Zhao, J.M. Kim, G. Stucky, H.J. Shin, R. Ryoo, Direct imaging of the pores and cages of three-dimensional mesoporous materials, Nature 408 (2000) 449–453.
[7] L. Han, Y. Sakamoto, O. Terasaki, Y. Li, S. Che, Synthesis of carboxylic group functionalized mesoporous silicas (CFMSs) with various structures, J. Mater. Chem. 17 (2007) 1216–1221.
[8] Y. Ma, L. Han, K. Miyasaka, P. Oleynikov, S. Che, O. Terasaki, Structural study of hexagonal close-packed silica mesoporous crystal, Chem. Mater. 25 (2013) 2184–2191.
[9] C.T. Kresge, M.E. Leonowicz, W.J. Roth, J.C. Vartuli, J.S. Beck, Ordered mesoporous molecular-sieves synthesized by a liquid–crystal template mechanism, Nature 359 (1992) 710–712.
[10] D. Zhao, J. Feng, Q. Huo, N. Melosh, G.H. Fredrickson, B.F. Chmelka, G.D. Stucky, Triblock copolymer syntheses of mesoporous silica with periodic 50 to 300 angstrom pores, Science 279 (1998) 548–552.
[11] C. Jin, L. Han, S. Che, Synthesis of a DNA–silica complex with rare two-dimensional square p4mm symmetry, Angew. Chem. Int. Ed. Engl. 48 (2009) 9268–9272.
[12] S. Che, Z. Liu, T. Ohsuna, K. Sakamoto, O. Terasaki, T. Tatsumi, Synthesis and characterization of chiral mesoporous silica, Nature 429 (2004) 281–284.
[13] A.E. Garcia-Bennett, K. Miyasaka, O. Terasaki, S.N. Che, Structural solution of mesocaged material AMS-8, Chem. Mater. 16 (2004) 3597–3605.
[14] A.E. Garcia-Bennett, N. Kupferschmidt, Y. Sakamoto, S. Che, O. Terasaki, Synthesis of mesocage structures by kinetic control of self-assembly in anionic surfactants, Angew. Chem. Int. Ed. 44 (2005) 5317–5322.
[15] L. Han, K. Miyasaka, O. Terasaki, S. Che, Evolution of packing parameters in the structural changes of silica mesoporous crystals: cage-type, 2D cylindrical, bicontinuous diamond and gyroid, and lamellar, J. Am. Chem. Soc. 133 (2011) 11524–11533.

[16] X. Cao, D. Xu, Y. Yao, L. Han, O. Terasaki, S. Che, Interconversion of triply periodic constant mean curvature surface structures: from double diamond to single gyroid, Chem. Mater. 28 (2016) 3691–3702.
[17] C. Gao, Y. Sakamoto, K. Sakamoto, O. Terasaki, S. Che, Synthesis and characterization of mesoporous silica AMS-10 with bicontinuous cubic Pn3m symmetry, Angew. Chem. Int. Ed. 45 (2006) 4295–4298.
[18] L. Han, D. Xu, Y. Liu, T. Ohsuna, Y. Yao, C. Jiang, Y. Mai, Y. Cao, Y. Duan, S. Che, Synthesis and characterization of macroporous photonic structure that consists of azimuthally shifted double-diamond silica frameworks, Chem. Mater. 26 (2014) 7020–7028.
[19] W. Mao, X. Cao, Q. Sheng, L. Han, S. Che, Silica scaffold with shifted "Plumber's nightmare" networks and their interconversion into diamond networks, Angew. Chem. Int. Ed. 56 (2017) 10670–10675.
[20] Y. Han, D. Zhang, L.L. Chng, J. Sun, L. Zhao, X. Zou, J.Y. Ying, A tri-continuous mesoporous material with a silica pore wall following a hexagonal minimal surface, Nat. Chem. 1 (2009) 123–127.
[21] C. Gao, Y. Sakamoto, O. Terasaki, S. Che, Formation of diverse mesophases templated by a diprotic anionic surfactant, Chemistry 14 (2008) 11423–11428.
[22] C.H. Xiao, N. Fujita, K. Miyasaka, Y. Sakamoto, O. Terasaki, Dodecagonal tiling in mesoporous silica, Nature 487 (2012) 349–353.
[23] C.B. Gao, H.B. Qiu, W. Zeng, Y. Sakamoto, O. Terasaki, K. Sakamoto, Q. Chen, S.A. Che, Formation mechanism of anionic surfactant-templated mesoporous silica, Chem. Mater. 18 (2006) 3904–3914.
[24] M. Suga, S. Asahina, Y. Sakuda, H. Kazumori, H. Nishiyama, T. Nokuo, V. Alfredsson, T. Kjellman, S.M. Stevens, H.S. Cho, M. Cho, L. Han, S.N. Che, M.W. Anderson, F. Schuth, H.X. Deng, O.M. Yaghi, Z. Liu, H.Y. Jeong, A. Stein, K. Sakamoto, R. Ryoo, O. Terasaki, Recent progress in scanning electron microscopy for the characterization of fine structural details of nano materials, Prog. Solid State Chem. 42 (2014) 1–21.
[25] M. Kobayashi, K. Susuki, H. Otsuji, Y. Sakuda, S. Asahina, N. Kikuchi, T. Kanazawa, Y. Kuroda, H. Wada, A. Shimojima, K. Kuroda, Direct observation of the outermost surfaces of mesoporous silica thin films by high resolution ultralow voltage scanning electron microscopy, Langmuir 33 (2017) 2148–2156.
[26] K. Miyasaka, H. Hano, Y. Kubota, Y.Z. Lin, R. Ryoo, M. Takata, S. Kitagawa, A.V. Neimark, O. Terasaki, A stand-alone mesoporous crystal structure model from in situ X-ray diffraction: nitrogen adsorption on 3D cagelike mesoporous silica SBA-16, Chem. Eur. J. 18 (2012) 10300–10311.
[27] Y. Sakamoto, T.W. Kim, R. Ryoo, O. Terasaki, Three-dimensional structure of large-pore mesoporous cubic Ia(3)over-bard silica with complementary pores and its carbon replica by electron crystallography, Angew. Chem. Int. Ed. 43 (2004) 5231–5234.
[28] C.B. Gao, I. Izquierdo-Barba, I. Nakase, S. Futaki, J.F. Ruan, K. Sakamoto, Y. Sakamoto, K. Kuroda, O. Terasaki, S. Che, Mesostructured silica based delivery system for a drug with a peptide as a cell-penetrating vector, Microporous Mesoporous Mater. 122 (2009) 201–207.

CHAPTER THREE

Stimuli-Responsive Nanomachines and Caps for Drug Delivery

Navnita Kumar, Wei Chen, Chi-An Cheng, Tian Deng, Ruining Wang, Jeffrey I. Zink[1]

Department of Chemistry and Biochemistry, University of California, Los Angeles, CA, United States
[1]Corresponding author: e-mail address: zink@chem.ucla.edu

Contents

Abstract

In this review we focus on methods that are used to trap and release on command therapeutic drugs from mesoporous silica nanoparticles (MSNs). The pores in the MSNs are large enough to accommodate a wide range of cargo molecules such as anticancer and antibiotic drugs and yet small enough to be blocked by a variety of bulky molecules that act as caps. The caps are designed to be tightly attached to the pore openings and trap the cargo molecules without leakage, but upon application of a designed stimulus detach from the nanoparticles and release the cargo. Of special emphasis in this review are nanomachines that respond to stimuli administered from external sources such as light or magnetic fields, or from chemical stimuli produced by the biological system such as a general change in pH or redox potential, or a highly specific chemical produced by a cancer cell or infectious bacterium. The goal is to release a high local concentration of the cargo only where and when it is needed, thus minimizing off-target side effects. We discuss sophisticated reversible nanomachines but also discuss some useful caps that simply break off from the nanoparticles in response to the selected

The Enzymes, Volume 43
ISSN 1874-6047
https://doi.org/10.1016/bs.enz.2018.07.003

stimulus. Many ingenious systems have been and are being designed; we primarily highlight those that have been demonstrated to operate *in vitro* and/or *in vivo*. In most cases the closed MSNs are endocytosed by diseased or infected cells and opened inside the cells to release the drugs. We begin with an overview of the nanoparticles and nanomachines and then present examples of drug release triggered by internal chemical stimuli from the organism and finally by external light and magnetic field stimuli.

1. INTRODUCTION AND BACKGROUND

Major objectives in nanomedicine and nanotherapy include: the ability to trap therapeutic molecules inside of nano-carriers; carry therapeutics to the site of the disease with no leakage; release high local concentrations of drug; release only on demand—either autonomous or external; and kill the cancer or an infectious organism. In this review we focus on mesoporous silica nanoparticles (MSNs) as the nano-carriers and nanomachines or related capping agents to trap, carry and release therapeutic molecules. Localized release of high concentrations of therapeutics produce low concentrations when diluted in the total volume of blood and reduce off-target side effects. Nanomachines that carry out the stimulus-responsive local release of drugs consist of a solid support (the MSN), and moving parts (molecules that undergo large amplitude motion) that respond to either external spatial and temporal control, or to internal chemical stimuli specific to the site of the disease.

The mesoporous silica nanoparticles that are the backbone of the drug delivery platforms described in this review are the result of two key discoveries: the sol–gel synthesis of silica, and self-assembling surfactant templates of the pores. Silica, SiO_2, is one of the most abundant minerals on earth. A solution-based synthesis was first reported in 1847 by Ebelmen who discovered that alkoxysilanes formed an amorphous form of silica when exposed to moisture [1]. This reaction is now known as a sol–gel synthesis [2]. The initially formed gel can be dried to produce a xerogel under mild conditions, and organic or biomolecules that are dissolved in the initial solution can be trapped in the forming glass [3]. Trapped enzymes can be stabilized in the amorphous glass, retain their enzymatic reactive properties, and be used as biosensors [4,5].

The second key discovery, independently discovered by several groups, was that liquid crystalline phases of surfactant molecules can be used as templates to form highly structured pores in sol–gel glass. The structure that is primarily discussed in this review involves rod-shaped surfactant structures that are encapsulated in glass and upon removal by extraction or calcination leave behind empty tubular pores. The most famous example involved a cetylammonium halide template (CTAB) that was first reported in the patent literature in 1971, independently reported in 1990, then again in 1991 by Beck et al. and now known as "MCM41" [6–8]. The as-synthesized materials were multi-sized aggregates, but further refinements by many groups ultimately led to controllable synthesis of both nanofilms [9–13] and nanospheres. The latter structures containing mesopores (between 2 and 8 nm) are shown in Fig. 1 and discussed in this review.

MSNs have remarkable properties. Typical particles discussed in this review are spheres about 100 nm in diameter with tubular pores about 2.5 nm in diameter. One gram of particles has a surface area of about 1000 m^2 and a pore volume of about 1 cm^3. Because typical anticancer drugs have a density of about 1 g/cm^3, the carrying capacity can be 100 wt% (although this upper limit is rarely achieved because of imperfect molecular packing).

Are mesostructured silica nanoparticles safe? The answer is generally yes. It is beyond the scope of this article to go into detail, but it is important to note that "not all silica is created equal"; some forms such as "fumed" silica made at very high temperatures have very different properties from those of solution-based sol–gel silica [2,14]. Many *in vivo* studies have been carried out with the latter with no adverse effects from the nanoparticles.

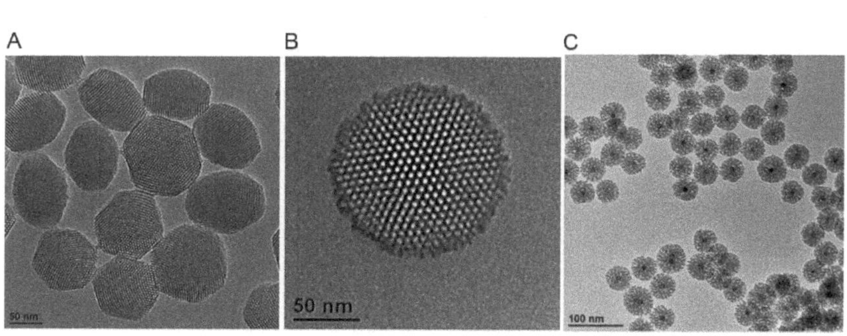

Fig. 1 TEM images of (A) typical mesoporous silica nanoparticles, (B) large pore silica nanoparticles, and (C) core@shell nanoparticles.

It is also important to note that silica dissolves in water; ocean water contains enough silicic acid that diatoms form bio-mineralized silica structures. Silica is excreted by mammals in the feces and urine.

Methods of chemically derivatizing silica surfaces are well established and can be applied to MSNs. The molecules to be attached can be added to the initial preparation solution (co-synthesis) or after formation of the MSNs (post-synthesis grafting). Pure MSNs aggregate in aqueous solution, but chemically attaching charged molecules to the outer surfaces stabilizes colloidal suspensions. Fluorescent molecules are attached in order to use the particles for imaging applications, and biomolecules are attached to actively target cancer cells.

The diverse capping systems can be broadly categorized into three main groups as seen in Fig. 2. Reusable caps are based on reversible binding constants of a bulky capping molecule. Completely reversible caps work on the principle of reversal of affinity of a bulky ring shaped molecule threaded on a stalk containing two or more binding sites after stimulation. Irreversible caps function via chemical bond cleavage of the bulky capping molecule, thus permanently detaching them from the nanomachines.

Stimuli-responsive drug delivery in general, and nanomachines in particular, require a biocompatible stimulus and/or power supply. How can the delivery system be activated when the particles are deep in the body? Two categories of stimuli are being used: external and autonomous. External activation has the desirable property of being under control by the operator; it can be turned on or off as desired and it can be localized to the site of the disease. Examples using external activation by light and magnetic fields will be presented. Magnetic fields penetrate tissue far better than light, but the latter can be tightly focused. The second category, autonomous activation, takes advantage of chemical properties that are as specific as possible to the site of the disease. The most widely used properties include pH change (from pH 7.4 in blood to 6 or below in tumors or lysosomes of cells), redox change (antioxidants such as glutathione in the cytosols of cells), enzymes over expressed by tumor cells or antigens, or antibodies specific to tumor cells. These subjects have been reviewed extensively [15–25]. In the following sections, each of these types of stimuli will be discussed in conjunction with different types of nanomachines and related capping systems. In this review, the results of attaching nanomachines to the particles in order to control the pore openings and trap and release therapeutic drugs in response to designed stimuli will be discussed.

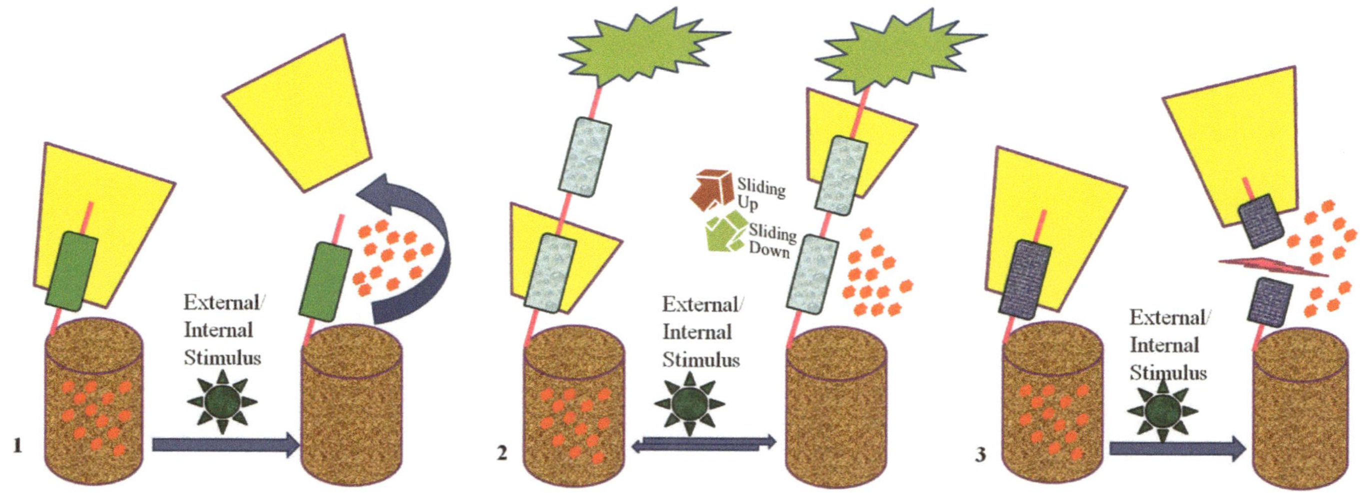

Fig. 2 Sketch of the three major modes of operation of pore uncapping. (1) Reusable valve where a bulky cyclic molecule has a binding constant to the attached thread-like "stalk" that decreases when stimulated. (2) Reversible valve where the bulky cyclic molecule has a binding constant at a position close to the pore opening that decreases allowing it to slide away from the pore opening but not leave the stalk. (3) A bulky molecule is chemically bonded to the pore opening and the bond is broken when stimulated.

2. INTERNAL/AUTONOMOUS STIMULI-RESPONSIVE DRUG DELIVERY

The bulk of the nano-carriers reported in the recent literature are ones that respond to changes in pH or redox potential in the interiors of cells such as lowering of pH on endocytosis or changes of redox potential between the extracellular environment and the cytosol after cellular uptake of the nano-carriers.

2.1 pH-Sensitive Nano-Carriers

The pore opening depends on structural changes caused by changing the interactions of the capping system with other charged species (stalk or the pore orifice) due to proton addition or abstraction. Thus functionalization of these nanomachines by groups that can be hydrolyzed by pH, or coating them with polymers, that can be dissolved with changes in pH is a great way of using pH as an internal stimulus. Nanomachines that are responsive to a change in pH from 7.4 (blood) to pH 6 and lower (lysosomes and tumor interiors) have been designed and tested *in vitro* and *in vivo*. A very successful example is based on supramolecular rotaxanes and pseudorotaxanes. This design consists of a stalk that is bonded to the MSN pore openings and a cyclic molecule that threads onto the stalk and blocks the pore. The non-covalent binding is caused by hydrophobic interactions between the stalk and ring. The stalk is designed such that at pH values lower than 7 the stalk is protonated and becomes hydrophilic thus decreasing the interaction strength and allowing the ring to escape thus opening the valve. This design is reusable upon increasing the pH. A completely reversible design uses a stalk with two different binding sites with the most hydrophobic closest to the pore opening. Upon acidification the ring moves away to the more distant binding site thus opening the valve. It moves back when the pH is increased.

2.1.1 Reusable Valve

A simple elegant reusable valve (as seen in Fig. 3) has been used to deliver therapeutic drugs *in vitro* and *in vivo*. An important nanovalve was reported by Meng et al. showing *in vitro* operation of an autonomously controlled nanovalve opening inside cells of interest [26]. This nanovalve was synthesized using a benzimidazole based silane (stalk) which was chemically conjugated to the MSN surface through silanol condensation and then

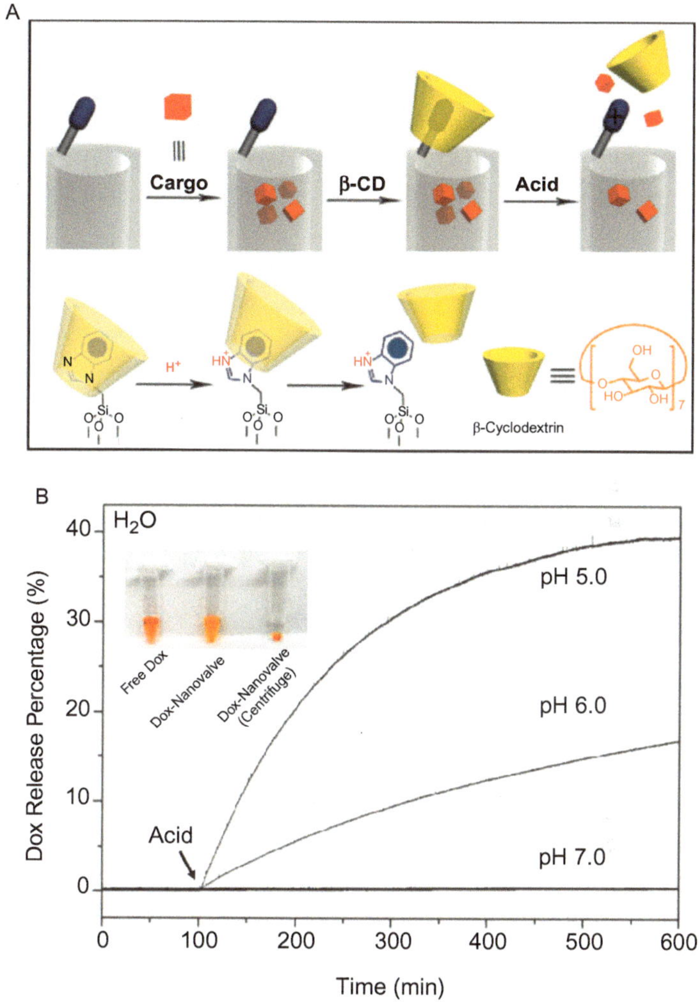

Fig. 3 (A) An example of a reusable cap where the hydrophobic interior of cyclodextrin is attracted to the hydrophobic part of the stalk at pH 7.4. Acidification protonates the stalk causing it to become hydrophilic thus releasing the cyclodextrin. (B) Release profiles of doxorubicin. *Adapted with permission from H. Meng, M. Xue, T. Xia, Y.-L. Zhao, F. Tamanoi, J.F. Stoddart, J.I. Zink, A.E. Nel, Autonomous in vitro anticancer drug release from mesoporous Silica nanoparticles by pH-sensitive nanovalves, J. Am. Chem. Soc. 132 (2010) 12690–12697. Copyright (2010) American Chemical Society.*

capping by beta-cyclodextrin to form a physical blockage over the pore openings, preventing leakage of the guest molecules (drug molecules) from the MSN at physiological pH (~7). The chemical structure of the stalk component was carefully chosen and modified to realize the operation of the nanovalve under the desired pH conditions. The cap was released from the stalk at endosomal acidification conditions in human differentiated myeloid (THP-1) and squamous carcinoma (KB31) cell lines. The authors also carried a systematic study of the surface functionalization of the MSNP to optimize an effective and rapid *in vitro* release of the anticancer drug doxorubicin to the nuclei of KB-31 cells. The release behavior monitored by fluorescence spectroscopy clearly indicated a direct relationship between the pore opening event and the pH of the surroundings. The *in vitro* studies demonstrated cell killing by the release of doxorubicin within the acidic cellular environment. Importantly, with the addition of NH_4Cl that inhibited the lysosomal acidification, no killing occurred proving that the system was indeed pH-responsive.

In a study involving tularemia, an intracellular infectious disease, the drug moxifloxacin was successfully delivered both *in vitro* and *in vivo* using the same nanomachine. The interior of the nanoparticle was modified with phosphonate silanes to increase the loading capacity of moxifloxacin. The *in vivo* efficacy of this pH-responsive delivery vehicle was 2.7-fold better than same amount of free drug [27,28].

2.1.2 Reversible Valves

An example of a completely reversible valve has been reported by Angelos et al. The molecular machines are composed of cucurbit(6)uril [CB(6)] rings that encircle tethered trisammonium stalks. The rings move between two sites under different pH conditions depending on the pK_a values of the binding sites. This pH operated "dialed-in" clock system is a novel entity due to the selective switch operation to release the cargo at a low pH as well as at a high pH [29]. At neutral pH, the CB(6) ring surrounds the tetramethylenediammonium recognition unit, blocking the nanopores. However, at low pH (~5) the aniline nitrogen atom is protonated thus moving the CB(6) ring toward the end of the stalk containing a hexamethylenediammonium station and allowing the cargo to come out of the pores. At high pH (~10) the nitrogen atoms on the stalk are deprotonated thus dethreading the CB(6) ring and releasing the cargo. The authors also showed that a selective chemical modification of the stalk can fine-tune the system thus resulting in cargo release at a desired pH.

2.1.3 Swinging Gates

Another type of pH based nanomachine called a swinging gate uses a different design strategy. The authors used reversible coordination complex

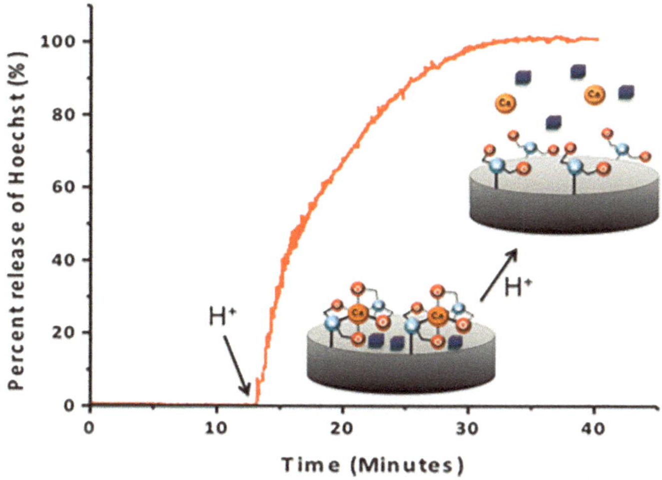

Fig. 4 Two chelating ligands attached to the pore opening bind to a metal ion thus latching them together over the pore and capping it. Upon acidification the metal-ligand binding decreases releasing the metal ion and unblocking the pore. *Adapted with permission from D. Tarn, C.-J. Yu, J. Lu, A. Hartz, F. Tamanoi, J.I. Zink,* In vitro *delivery of calcium ions by nanogated mesoporous silica nanoparticles to induce cancer cellular apoptosis, Mol. Sys. Des. Eng. 2 (2017) 384–392. Copyright (2017) Royal Society of Chemistry.*

formation between the metal ions and the ligands to devise a nanogate as seen in Fig. 4. Two ligand molecules that can bind metal ions are bonded on opposite sides of a pore opening. In the presence of a metal ion at pH >7, the molecules stretch across the pore opening to bind the metal and thus block the pore. Upon acidification the binding constant decreases (in part due to protonation of the Lewis bases), the metal ion is released and the pore is opened releasing the cargo molecules in the pores. Based on the above theme, a nano-carrier was designed by Tarn et al. [30]. The pore openings on the mesoporous silica nanoparticles were grafted with iminodiacetic acid (IDA) molecules. Two of the iminodiacetic acid (IDA) molecules are latched together by a metal ion forming a bis-IDA metal complex that acts as a closed gate. Hoechst 33342 dye was used as the cargo molecule and released by acidification because the metal-IDA binding constant decreased and unlatched the gate.

An alternative use of the swinging gate design is to deliver the metal ion itself. By using the swinging gates the problem of not being able to store the small metal ions in the pores of nanoparticles was overcome. *In vitro* delivery of calcium ions to cancer cells is an important bio-application of this kind of system [31]. Delivery of calcium to the cell cytoplasm to induce cellular death is an attractive alternative to the traditional and more hazardous chemotherapy drugs. Treatment of MiaPaCa-2 cancer cells with Hoechst-loaded, calcium-latched nanogate-modified MSNs resulted in the autonomous release of calcium ions as visualized by cell nuclei staining and calcium-induced apoptosis. Apoptosis was confirmed by observation of nuclear fragmentation, chromatin condensation, and Western blotting for cleaved caspase-3. Sometime the metal latched nanogates cannot completely prevent the cargo from leaching out. Du et al. have overcome this difficulty by addition of myoglobin, a protein with five histidine residues on its outer surface, completely capping the pores after the coordination of the ligands attached at the pore orifice with the Cu^{2+} cation [32].

2.1.4 Molecular Recognition Based Gates

A second type of gating mechanism uses molecular recognition and hydrogen bonding. An ingenious pH mediated nanomachine was synthesized by Theron et al. utilizing the principle of molecular recognition similar to that used in the case of DNA formation by base pairing using hydrogen bonding [33]. The authors surface-functionalized MSNs using a tri-azine or uracil fragment. After loading with cargo molecules, the nanoparticles were successfully capped through hydrogen bonds with the complementary caps linked to bulky cyclodextrin rings. *In vitro* studies demonstrated no leakage of cargo at neutral pH and a pH-stimulated release at acidic pH. The delivery and release of the anticancer drug camptothecin (CPT) was also tested. A very low dose of CPT (40 times less than what was used for free CPT) was as efficient at cell killing as free CPT. Another nano-carrier based on a similar molecular recognition principle was also developed by the same group. This time they used inexpensive and commercially available cyanuric acid as a stopper instead of the bulky cyclodextrin-attached complementary counterpart [34]. This modification not only was simpler to synthesize but also involved more molecular recognition sites and showed a high efficiency of drug CPT delivery for colon cancer therapy. The *in vitro* experiments on the HCT-116 cell line were congruent with *in vitro* experiments done on primary cultures from patient tumor samples. These nano-carriers show strong internalization into the cancer cells and then the cargo release by the pH stimulation.

2.1.5 Nano Pistons

A pH mediated dual cargo system was designed by Zhao et al. where cyclodextrin rings were chemically bonded over the openings of the nanopores and conjugated with plugs like rhodamine B/benzidine forming a nanopiston that shows a dynamic movement in and out of the cylindrical cavity [35]. The sizes of the openings of the nanopores are defined by the uniform diameter of the cyclodextrin rings. The cyclodextrin further helps in loading of different cargos by hydrophobic interactions. The molecular plugs and the cyclodextrin rings can be cleaved at different pH thus releasing small and large drug molecules, respectively.

2.1.6 Megagates

Another cargo size-selective pH-responsive megagate system was reported by Xue et al. [36]. This system was based on the concept that even partial closure of the large pores can successfully trap big cargos. These megagates were assembled around mesoporous silica material SBA-15 to mechanize large pores (diameter ~6.5 nm). This nanosystem has an on demand release in a very precise and controlled manner by changing the number and lengths of the arms of the megagates. The megagate on SBA-15 is capable of storing disparate sized cargos and then releasing them upon pH stimulation.

2.1.7 Polymer-Based Gates

A variety of acid-responsive polymers have been used as capping agents. A polyamine based gating system was developed where at acidic pH the amines are protonated thus forming a rigid structure and blocking the pores. At neutral pH the amines are deprotonated thus breaking the rigid structure and opening the pores and releasing the cargos [37–39]. In another example, the pores are functionalized with amino groups. At an acidic pH, the polymer caps the pores via the electrostatic interactions between the protonated amino groups and its sulphonyl groups. At neutral pH, deprotonation of the amino groups eliminates the attraction and releases the cargo [40]. Chan et al. have functionalized their core@shell nanoparticles with acid-labile imine bonds using APTES and ferrocene carboxaldehyde. Breaking the bonds at low pH releases the cargo which can be easily stored inside at a neutral pH [41].

Coordination polymers have also been used as capping systems [42]. Che et al. functionalized the surface of the MSNs with aminopropyl moieties and loaded them with topoisomerase I inhibitor topotecan at neutral pH. The system was capped by introducing Zn^{2+} cations and 1,4-bis (imidazol-1-ylmethyl) benzene forming a coordination polymer shell around the MSNs. At lower pH, amines of the aminopropyl moieties are protonated thus disintegrating the whole coordination shell. There are several polymers that act as capping agents [43–52], via size modulation with change in pH. For example, Sailor et al. used a porous silica film loaded with insulin and functionalized it with a chitosan based hydrogel through a cross-linking reaction between chitosan and GTPMS [43]. At pH 6, the hydrogel layer swells releasing the insulin whereas at neutral pH chitosan adheres to the nanoparticle thus trapping the insulin within it. Using the layer-by-layer (LbL) deposition technique, several pH-triggered capping systems on MSNs have been developed [53–55]. In this technique, there is deposition of alternating layers of opposite charges. Depending on the interaction of cargos with the polymeric layer, the cargo release can either be achieved at neutral pH while loading at acidic pH or vice versa. For instance Zhu et al. [53] coated DOX loaded silica nanotubes with sodium polystyrene sulfonate (PSS) and PAH. The nanosystem was capped at a basic pH and cargo was delivered at acidic pH. In contrast, the same group devised similar nanotubes coated with alginate and chitosan after surface modification with APTES. This nanosystem remained capped at a pH $\sim$4 while the fluorescein cargo was delivered at basic pH.

2.2 Redox-Sensitive Nano-Carriers

Reduction and oxidation reactions (Redox) provide a second important stimulus for nanomachines. For biological applications the increased level of reducing agents (*e.g.*, glutathione) within the cellular environment over that in the bloodstream can be used to activate redox-responsive nanovalves. Historically, both oxidation and reduction reactions were used to stimulate the first supramolecular nanovalves on silica. For biological applications, the simplest and most versatile pore-capping devices are based on disulfide bonds that are cleaved by reducing agents. These systems are described in this section.

2.2.1 Rotaxane Based Redox-Nanovalves

The first supramolecular nanomachine operating on silica was based on bistable [2]rotaxanes attached onto the silica framework [56]. The cyclic molecule cyclobis-(paraquat-*p*-phenylene) (CBPQT4+) could shuttle between tetrathiafulvalene (TTF) and 1,5-dioxynaphthalene (DNP) units on a stalk under different redox conditions. This thread/dethread behavior resulted in an on/off position of this supermolecular nanovalve to trap cargo molecules within the pores and then release them. Both reusable (opening only) and totally reversible examples operated by sequentially adding oxidizing and reducing agents were synthesized [57,58]. These pioneering studies on a fine-tuned redox-responsive nanovalve paved the foundation for future design of drug delivery systems even though the chemicals and solvents used for this system are not bio-relevant.

2.2.2 Disulfide Based Redox-Nanovalves

The most versatile redox-operated capping system is based on bulky groups attached to MSNs by linkers containing the disulfide -S-S- linkage. When attached to pore openings the bulky groups block the pore openings but exposure to reducing agents cleaves the disulfide and releases the bulky group and the cargo. The first example of this strategy was reported by Lin et al. who utilized inorganic CdS nanocrystals as the bulky gatekeeper which was conjugated to MSNs through the disulfide bond [59]. The average size of the CdS nanocrystal was around 2 nm, suitable for blocking the pore opening of MSNs. Upon the addition of mercaptoethanol (ME), the cargo was released. Subsequently, a multitude of different bulky groups have been used. Three bio-relevant examples are (a) beta-cyclodextrin associated with adamantane, (b) cucurbit(6)uril associated with a diamine stalk, and (c) an oligopeptide holding an alpha-cyclodextrin in place as the bulky capping agent. Upon the addition of a reducing agent, the disulfide bond was cleaved inducing an open access of the mesoporous and cargo release [60].

The beta-cyclodextrin associated with adamantane system (as seen in Fig. 5) was used in a detailed *in vivo* study of treatment for tularemia, a disease caused by the biowarfare bacterium *Francisella tularensis*[61]. This intracellular infectious disease is caused by bacteria inside of white blood cells. The antibiotic cargo moxifloxacin was carried into the cytosol of the infected cells, released by reduction of disulfide linkages by glutathione, and killed the bacteria. In mouse model studies this redox-responsive delivery system showed superior efficacy up to fivefold over a same amount dose of free drug.

2.2.3 Ferrocene-Based Redox-Nanovalves

A few other redox-activation motifs [62] have been reported (primarily using ferrocene linkages [63,64]) but this area is underdeveloped. In one such study, Khashab et al. utilized host–guest interactions between ferrocene beta-cyclodextrin as a capping system on MSNs. Ferrocenedicarboxylic acid was conjugated to the MSN surface via an amine-silane linker. β-CD was then introduced to bind with the ferrocene for capping. By electrochemically oxidizing ferrocene the β-CD dissociated and cargo was released.

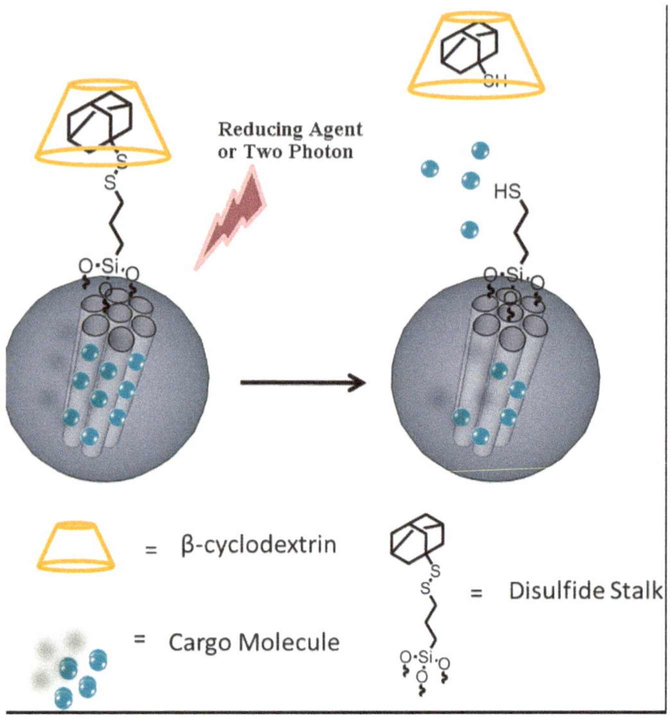

Fig. 5 Redox activation of a capping system consisting of a bulky cyclodextrin attached to the pore opening by a disulfide bond. Reduction (by glutathione or by photo-induced electron transfer, Section 3.1.5) breaks the bond and releases the cap. *Adapted with permission from T.M. Guardado-Alvarez, L.S. Devi, J.-M. Vabre, T.A. Pecorelli, B.J. Schwartz, J.-O. Durand, O. Mongin, M. Blanchard-Desce, J.I. Zink, Photo-redox activated drug delivery systems operating under two photon excitation in the near-IR, Nanoscale 6 (2014) 4652–4658. Copyright (2014) Royal Society of Chemistry.*

Even though a test release experiment was conducted in an electrochemical cell with little bio-relevance, the ferrocene-based redox-responsive system still opens up new directions to novel nanomachines.

2.3 Enzyme Triggered Delivery System

Two general types of designs have been used to cap pores and then open them in response to enzymatic activity. The first and most machine-like were called "snap-tops." As is the case with nanovalves, a cyclic molecule on a stalk is used to block the pore. To keep the cyclic molecule in place, a large "stopper" molecule is attached to the end of the stalk with functional groups that are cleaved by enzymes. The closed system is snapped open when the enzyme breaks off the bulky stopper and the cyclic molecule moves away from the pore. The second design simply uses a long chain-like molecule attached to the pore openings. Enzymatic activity breaks the long molecules into pieces thus unblocking the pores. The ultimate goal is to use systems responsive only to enzymes overexpressed by specific (diseased) cells.

2.3.1 Snap-top Systems

An enzyme-responsive snap-top nanovalve based on (2)rotaxane and α-cyclodextrin (α-CD) (as seen in Fig. 6) was synthesized and snapped open to allow cargo release when porcine liver esterase (PLE) was added [65]. After cargo loading, α-CD was threaded to a tri(ethylene glycol) chain attached to MSNs. The stopper group was attached to the tri(ethylene glycol) chain by an ester linkage. PLE can achieve the hydrolysis of the adamantyl ester stopper followed by α-CD dethreading. Rhodamine B was used as a model drug, and the triggered release was monitored by luminescence spectroscopy. Dye was released only after adding PLE.

A similar design was used by He et al. In this design cathepsin B, an overexpressed protease in various tumors was selected to be the trigger

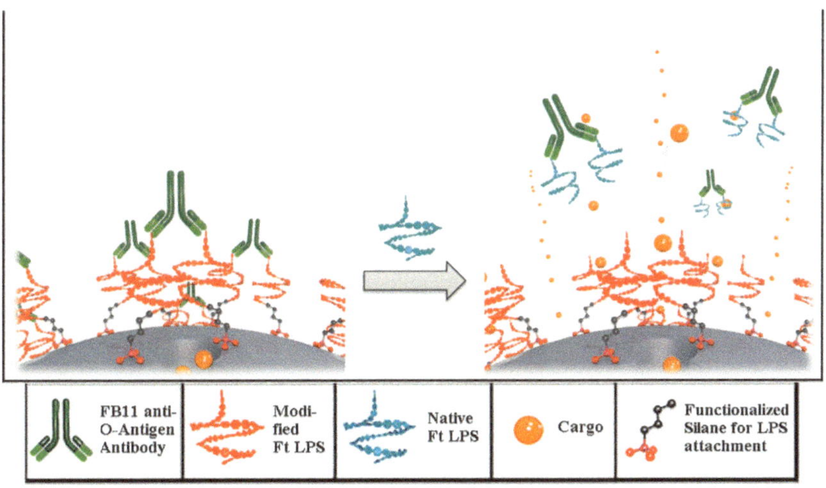

Fig. 6 An example of a "snap-top." A bulky cyclodextrin molecule is held in place by a stopper bonded to the end of the stalk by an ester linkage accessible to an esterase enzyme. When exposed to the enzyme the bond is hydrolyzed releasing the cyclodextrin cap and opening the pore. *Adapted with permission from K. Patel, S. Angelos, W.R. Dichtel, A. Coskun, Y.-W. Yang, J.I. Zink, J.F. Stoddart, Enzyme-responsive snap-top covered silica nanocontainers, J. Am. Chem. Soc. 130 (2008) 2382–2383. Copyright (2008) American Chemical Society.*

enzyme [66]. As before, an alkoxysilane tether and α-CD were used as the thread and capping agent, and serial peptide sequences were attached as a stopper. The stoppers included Gly-Phe-Leu-Gly (GFLG), a peptide that can be hydrolyzed by cathepsin B, and then it was followed by a cell-penetrating peptide of seven arginine (R7) sequence, and a tumor-targeting peptide of Arg-Gly-Asp-Ser (RGDS). The release behavior of doxorubicin (DOX) loaded MSN was studied in PBS buffer, and 80% of loaded DOX was released while only 4.3% from the control group without cathepsin B was delivered. This multifunctional MSN targeted and killed cancer cells *in vitro*.

2.3.2 Polymer-Based Systems

Other enzyme-responsive drug delivery systems have been demonstrated using polysaccharides and peptides. Polysaccharide such as guar gum is cleaved by an enzyme mixture from colonic microflora [67]. Peptides such as protamine is cleaved by a proteolytic enzyme trypsin [68] and ε-poly-L-lysine is broken down by proteases from *Streptomyces griseus*[69]. These polymers (polysaccharide/peptides) are attached directly to MSN surfaces and act as capping molecules. All of them showed stimulated release *in vitro*, indicating its promising future application.

2.4 Antigen–Antibody Competitive Binding Delivery System

There are many papers about attaching antigens or antibodies on MSNs for targeting purposes such as cancer detection [70] and targeted drug delivery [71]. Their use as nanovalves on MSNs to stimulate release of cargo is rare. Recently, a highly pathogen-specific drug delivery system was reported by Ruehle et al. involving the antigen of a lipopolysaccharide (LPS) of *F. tularensis* (Ft) [72] (Fig. 7). The antigen–antibody interaction is utilized not only as recognition but also as a pathogen killing signal. This antigen-responsive nanovalve can be triggered only by a specific antigen or pathogen. The antibody, which is the gatekeeper, was bonded to the antigen derivative attached to MSNs. As the antigen derivative has weaker binding affinity to the antibody, the original antigen can compete against it effectively and displace the antibody, leading to gate opening and cargo release. The modified antigen was attached by binding one of its carboxylic groups to amine-modified MSNs via the EDC/NHS reaction and characterized by immuno-staining assay. To reduce the binding affinity of antibody and antigen on the MSN's surface, acetylation of OH groups on the antigen was carried out. The drug loading and release capacity was tested using fluorescein.

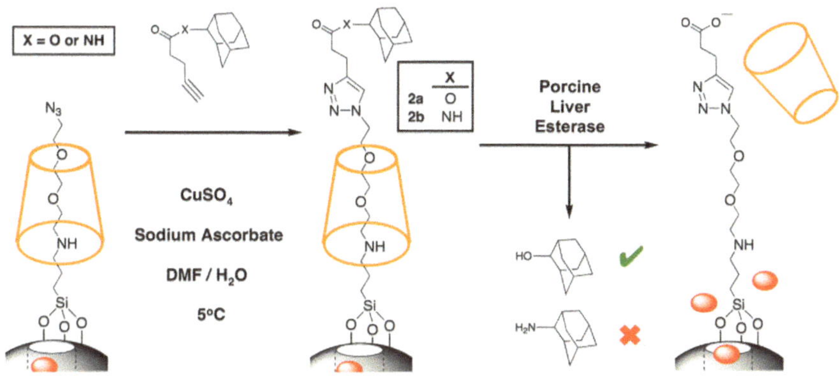

Fig. 7 An example of a competitive binding. An antibody on a synthetic antigen attached to the pore opening is displaced by the natural LPS antigen (green) from *Francisella tularensis* thus triggering the opening the pore and release of the cargo (orange spheres). *Adapted with permission from B. Ruehle, D.L. Clemens, B.-Y. Lee, M.A. Horwitz, J.I. Zink, A pathogen-specific cargo delivery platform based on mesoporous silica nanoparticles, J. Am. Chem. Soc. 139 (2017) 6663–6668. Copyright (2017) American Chemical Society.*

When the capped MSNs were incubated with Ft, the LPS from the bacteria displaced the antibody and the cargo was released. The release behavior was also studied with *Francisella novocida* (Fn), a bacterium with similar O-antigen structures, but no pronounced release was observed, illustrating the high selectivity of this drug delivery system *in vitro*. This system can recognize and signal the presence of a specific pathogen with high specificity, and when loaded with antibiotics will be able to kill them.

Similar to the antigen–antibody competitive binding capping system mentioned above, selective binding between Lewis X (Le^x), a glycan antigen, and *Aleuria aurantia* lectin (AAL), a fucose-specific carbohydrate-binding protein, was utilized for stimulated drug delivery [73]. Le^x was found to be overexpressed in cancer cells, and its antigenic epitope has been clinically used as a tumor marker for various malignancies such as pancreatic cancer, colorectal cancer and bladder cancer [74–76]. In this work, a Le^x antigen derivative attached on MSNs and capped with AAL. A dye (ATTO 430LS) was used as a model drug to test the delivery properties in DLD-1 human colorectal adenocarcinoma cell line where Le^x is overexpressed. When functionalized MSNs were incubated with colon cancer cells, the AAL cap will bind to Le^x overexpressed on the cells' surfaces and uncap the pores. Also, the Le^x antigen derivative on MSNs helped target cancer cells. The result of an *in vitro* study showed that the cell viability was not affected by functionalized MSNs, and that the dye emission was triggered after incubation with cancer cells and was threefold higher than the control group of cells without Le^x overexpression.

3. EXTERNAL STIMULI-RESPONSIVE DRUG DELIVERY

3.1 Light-Activated Nanomachines

Light-responsive drug delivery systems based on MSNs provide the advantage of remote control of the stimulated release. In addition, the high spatial resolution of photo-activation makes spatiotemporal release control feasible. Photo-activated drug delivery systems offer the potential of enhancing the treatment efficacy and reducing premature cargo release that can cause undesirable side effects. Mechanized MSNs that have photo-responsive cargo release properties are discussed in the following sections.

3.1.1 Impellers

Molecules that undergo photo-induced large amplitude motion but are motionless in the dark form the basis for nanomachines called impellers (as seen in Fig. 8). Azobenzene derivatives, for example, undergo photo-induced reversible *trans–cis* isomerization when the molecules are irradiated with blue light [77]. When these types of molecules are attached to the interior pore walls, unattached cargo molecules are trapped in the dark but when illuminated the continuous dynamic wagging motions of the phenyl rings release the cargo from nanoparticles' pores [78]. Angelos et al. chemically bonded azobenzene derivatives in the interiors MSNs to act as impellers [79]. When illuminated and excited at a wavelength at 457 nm, the continuous and reversible *trans–cis* photoisomerization of azobenzene occurred. During this process, the fluorescent molecules loaded in the pores of MSNs were continuously expelled from the mesopores. Such remotely controllable cargo release study by photo-activated manner opened up the possibilities of drug molecules loading and delivery into living cells on demand. Lu et al. loaded an anticancer drug, camptothecin (CPT), and other dye molecules in the azobenzene derivatized MSNs [80]. The results showed that the release of CPT and dye molecules from azobenzene derivatized MSNs in the cancer cells only occurred when the particles were illuminated with specific wavelength light that could initiate the photoisomerization. After photo-activation in the cancer cells, such nanoimpeller modified nanoparticles released CPT and induced the apoptosis of the cancer cells. Impellers were the first artificial nanomachine operated inside living cells. Further studies using continuous and real time monitoring of

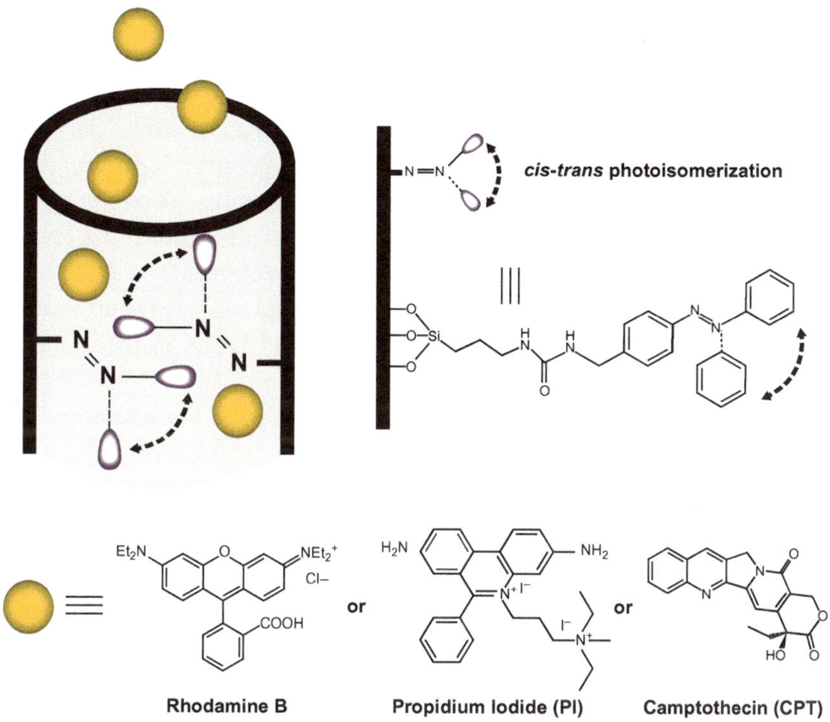

Fig. 8 An example of an impeller. Azobenzene molecules attached to the interiors of pores are static in the dark but undergo photo-induced *trans–cis* isomerization resulting in a large amplitude "wagging" motion. Cargo molecules are prevented from escaping from the pores in the dark but escape when the particles are irradiated. *Adapted with permission from J. Lu, E. Choi, F. Tamanoi, J.I. Zink, Light-activated nanoimpeller-controlled drug release in cancer cells, Small 4 (2008) 421–426. Copyright (2008) John Wiley and Sons.*

propidium iodide (PI), a cell membrane impermeable fluorescent dye released from the nanoimpeller systems in a single living cell were also carried out [81]. The real time spectroscopic measurement showed that after the photo-activation, PI molecules were released into cytosol and eventually intercalated with nuclear DNA based on the change of PI's fluorescence intensity and the wavelength shift.

3.1.2 Azobenzene/Cyclodextrin Valves

Azobenzene derivatives, in addition to acting as impellers, can bind cyclodextrin when in the *trans* configuration and de-bind in the *cis* configuration. When the azobenzene component is attached around the pore openings, the binding/debinding can serve as gatekeepers to control and regulate cargo release from MSNs. Ferris et al. attached two types of azobenzene derivatives that bind β-cyclodextrin (β-CD) onto the outer surface of MSNs [82].

When irradiated with UV light, the *trans* azobenzene photoisomerizes to the *cis* configuration, releases the β-CD cap, and causes cargo release from the mesopores. This type of gatekeeper system was modified by Tarn et al. [83] by attaching bulky stoppers to the ends of azobenzene derivatives that prevent α-CD from completely dissociating after light irradiation and the capping and uncapping becomes fully reversible. Wang et al. constructed nanovalves with tetra-ortho-methoxy-substituted azobenzene (mAzo) and β-CD on the surface of MSNs to control release of doxorubicin (DOX) from the nanopores. The red-light responsive property of this Azo gatekeeper system could induce the release of DOX in tissue [84].

Azobenzene derivatives modified on the surface of MSNs also conjugate with other functional molecules and act as gatekeepers. For example, Yuan et al. conjugated azobenzene modified double strand DNA on the surface of MSNs [85]. The DNA and azobenzene gatekeepers undergo photo-operated hybridization/dehybridization due to the photoisomerization of azobenzene and thus function as a reversible capping system. Recently, Hu et al. assembled iron oxide nanoparticles on the surface of hollow MSNs through host–guest complexation of cucurbit(8)uril and azobenzene [86]. The photo-activated dissociation of the iron oxide nanoparticles released the cargo molecules. In summary, versatile azobenzene derivatives can act as photo-activated nanovalves in addition to their function as impellers for therapeutic drug delivery.

3.1.3 Photo-Cleavage

Perhaps the simplest way to use light to release cargo molecules is through direct photo-cleavage of a capping molecule. Several examples have been demonstrated. Agostini et al. synthesized a methoxybenzylamine derivative molecule which was conjugated on the surface of MSNs [87]. After being illuminated with UV light, the photolytic bond breaking of the caps reduced the steric hindrance around the pore openings and led to the dye release. As another example, photo-labile coumarin derivatives bound with bulky β-CD were also designed and modified on the surface of MSNs [88]. Photolysis released the cap and thus the cargo. To achieve greater penetration depth, two-photon NIR excitation was used to cleave the photo-sensitive molecules. Such two-photon NIR triggered dye release from MSNs enables the photo-activated drug delivery from photo-cleavable gatekeeper systems in biological applications to be practical.

A different type of photoactivity was demonstrated using 2-nitrobenzyl conjugated poly(*N*-isopropylacrylamide) poly(NIPAMNBAE) attached on the surface of MSNs [89]. Poly(NIPAMNBAE) has a critical solution temperature (LCST) below 37 °C. Upon UV irradiation, the hydrophobic 2-nitrobenzyl acrylate was photolyzed resulting in an increase of the LCST of the polymer. Hence, the polymer changed to a coil conformation on the surface which allowed the cargo release from MSNs. The release of fluorescein molecules from poly(NIPAMNBAE) gated MSNs after the UV irradiation was successfully demonstrated in liver cancer cells.

3.1.4 Plasmonic Heating

Light-activated heating of gold nanoparticles has been used to trigger cargo release using thermo-sensitive gatekeeper-modified MSNs. For instance, Croissant et al. synthesized a gold core@MSN shell structure containing 20 nm gold nanoparticles as the cores and 150 nm mesoporous silica as the shell [90]. The thermal-sensitive nanovalves were constructed by capping and binding cucurbit(6)uril (CB6) on the stalks on the surface. The plasmonic heating generated from the gold core upon photo-activation raised the local temperature and decreased the binding constant between CB(6) and the stalk, thus unblocking the pores and releasing the cargo from MSNs. As another example, Li et al. constructed a mechanized hybrid system based on core@shell nanoparticles containing a gold nanorod as the core and mesoporous silica as the shell (Au@MSN) [91]. The thermal-sensitive gatekeepers modified on the surface were the host–guest complexation of sulfonatocalix(4)arene to the quaternary ammonium salt stalk that acts as the supramolecule switch. The binding constant of sulfonatocalix(4)arene and quaternary ammonium salt is temperature dependent. Upon exposure to NIR irradiation, the plasmonic heating generated from the gold nanorods in the mesoporous silica shell caused the dissociation of host–guest complexations on the surface and thus released the trapped cargo.

Thermo-sensitive phase-changing molecules were also designed as the gatekeepers to regulate the drug release from MSNs. For example, Li et al. introduced 1-tetradecanol, a biocompatible phase-changing molecule, as the gatekeeper to control doxorubicin (DOX) delivery from Au@MSN [92]. The melting temperature ($T_m = 39\,°C$) of the phase-changing molecules is higher than the physiological temperature. Upon either external heating from a water bath or the internal heating generated by plasmonic heating of the gold nanorods, DOX release from mesoporous shell was observed and was used to kill cancer cells. Such photo-activated drug delivery studies by the localized and internal heating make it possible for remote control of drug delivery in biological systems [93,94].

3.1.5 Transduced Photo-activation

Many of the nanomachines discussed in this chapter can be activated indirectly by light. The key is to use "transducer" molecules, i.e., molecules that after excitation can transfer protons, electrons or energy to the nanomachine. There are many known examples of photoacids and photo-reducing agents that can or have been attached to mechanized particles. An important subset of transducers are those that have a high cross-section (propensity) for absorbing two-photons simultaneously [95]. Their importance lies in their ability to be excited to an energy level twice that of the individual energies of the photons so that tissue penetrating infrared light can be used to drive molecular machines that do not absorb infrared light. Photoacids, photo-redox and energy transfer are discussed briefly in this section; two-photon photodynamic therapies are discussed in detail in chapter "Mesoporous silica-based nanoparticles for light-actuated biomedical applications via near-infrared two-photon absorption" by Croissant and Durand.

3.1.6 Photoacid Transducers

Photoacid molecules have pK_a values in their excited electronic states that are much lower than those in their ground states [96]. While the molecules are under irradiation, the proton transfer process decreases the pH value around the photoacid molecules. Such localized pH change can activate the capping and uncapping of mechanized MSNs. For example, Guardado-Alvarez et al. attached both photoacid molecules and pH-responsive nanovalves on the surface of MSNs [97]. When illuminated at 408 nm, the proton transfer from the photoacid molecules activated and opened the nanovalves followed by the cargo release. The photoacid and pH-sensitive gatekeepers were conjugated on the surface of MSNs through click reactions which homogeneously distributed the organic functional groups on the surface. The conjugation method provides an efficient photo-responsive DOX delivery strategy which could kill MCF-7 cancer cells [98].

3.1.7 Photo-redox Transducers

Photo transducer molecules that act as reducing agents in their excited electronic states provide the possibility of triggering cargo release from redox-sensitive gatekeeper systems [99]. For example, Guardado-Alvarez et al. attached both photo transducer molecules and disulfide bond bridged snap-top capping systems on the surface of MSNs [100]. After one- or two-photon excitation of the transducers, the inter-molecular electron transfer reduced the disulfide bond and caused the cleavage of snap-top capping molecules. The cargo loaded in MSNs could thus be released from the carriers after the trigger. Such photo-redox triggered cargo delivery system can also be applied to other redox potential-sensitive delivery systems [101].

3.1.8 Energy Transfer (FRET) Transducers

Photoexcited molecules can transfer energy between themselves by fluorescence resonance energy transfer (FRET) if they are close enough together. For example, impellers can be stimulated even if the azobenzene component itself does not absorb the light [102]. Thus molecules that have high two-photon cross sections attached to nanoparticles "upconvert" the energy of two infrared photons and then transfer it to azobenzene and deliver drugs.

3.2 Magnetic Field-Activated Nanomachines

Magnetic fields, specifically alternating magnetic fields (AMFs), are currently being used in nanomedicine for two purposes: hyperthermia and drug delivery. In both cases, the therapy relies on magnetic heating of nanoparticles and in many respects parallels therapy resulting from plasmonic heating discussed in Section 4. Magnetic fields have a distinct and powerful advantage over other external stimuli. They are not only temporally and spatially controllable but also deeply and harmlessly tissue penetrating.

The most important magnetic component in the nanoplatform is a superparamagnetic nanoparticle, most commonly FDA approved superparamagnetic iron oxide nanoparticles (SPIONs). When the SPION@MSNs are exposed to AMFs, localized heat will be generated from the internal SPION cores due to either Néel heating or Brownian heating; the former is related to the changing orientation of the magnetization, and the latter is related to the changing orientation of the particle itself [103]. SPIONs are effective agents for inducing cell and tumor killing by hyperthermia. For drug delivery, (SPION) cores are embedded in mesoporous silica forming a core–shell structure (denoted SPION@MSNs), and the pore openings of SPION@MSNs are decorated with thermo-sensitive nanovalves that act as gatekeepers. At body temperature the pores are capped, but upon exposure to AMFs the heat opens the gatekeepers thus releasing the drug.

The synergism between heating and drug delivery may be very useful. However, overheating portions of a tumor results in necrotic tissue and may release cancer cells from underheated regions into the blood. The amount of heat can be controlled such that only the local heating of a particle on the nanoscale can release drug without significantly raising the temperature of the surrounding tissue. Dong et al. designed a MSN that contained both an encapsulated nanothermometer and a SPION and demonstrated that the local internal temperature inside the Fe_3O_4@SiO_2 core–shell nanoparticle can increase by >20 °C over the temperature of the bulk solution after 5 min exposure to AMFs [104].

The design and therapeutic applications of AMF-responsive nanocarriers are in its infancy. In the following subsections, we give examples of nanovalves according to their mechanism of response to the magnetic field: nanovalve activation; thermolytic chemical bond breaking; DNA dehybridization; and phase-changing polymers.

3.2.1 *Nanovalve Activation*

The first study that used magnetic field actuation of a nanomachine to demonstrate external control of the release of therapeutics *in vitro* was published in 2010 by Thomas et al. [105]. The authors used a supramolecular valve comprised of an alkylammonium thread and a cucurbit(6)uril ring, where cucurbit(6)uril binds the thread attached to a core@shell mesoporous silica nanoparticle with a zinc-doped iron oxide nanocrystal core as seen in Fig. 9. Cucurbit(6)uril has a hydrophobic cavity which encircles the alkyl chain of the thread through non-covalent London force, ion–dipole interactions and hydrogen bonding. The authors demonstrated the release of Rhodamine B at a bulk temperature of 0 °C under an AMF, showing that the internal localized heating from the magnetic core instead of the bulk heating was responsible for activation. The authors then demonstrated magnetic field-activated release of doxorubicin (DOX) in the breast cancer cell line MDA-MB-231 and induced apoptosis.

3.2.2 *Chemical Bond Breaking: Thermolysis*

Saint-Cricq et al. designed a nanocapping system which is responsive to magnetic fields through chemical bond breaking. The authors used a PEG polymer containing two 4,4′-azobis(4-cyanovaleric acid) functional groups and attached both ends of the polymer to the SPION@MSNs thus capping the pores [106]. Azo compounds are well known to be thermally degradable, and heating breaks the polymer into three smaller pieces. The authors demonstrated stimulated release of rhodamine 6G with magnetic field actuation and showed that both the polymer-modified MSNs and the degradation products caused by magnetic heating exhibited no cytotoxicity toward fibroblasts.

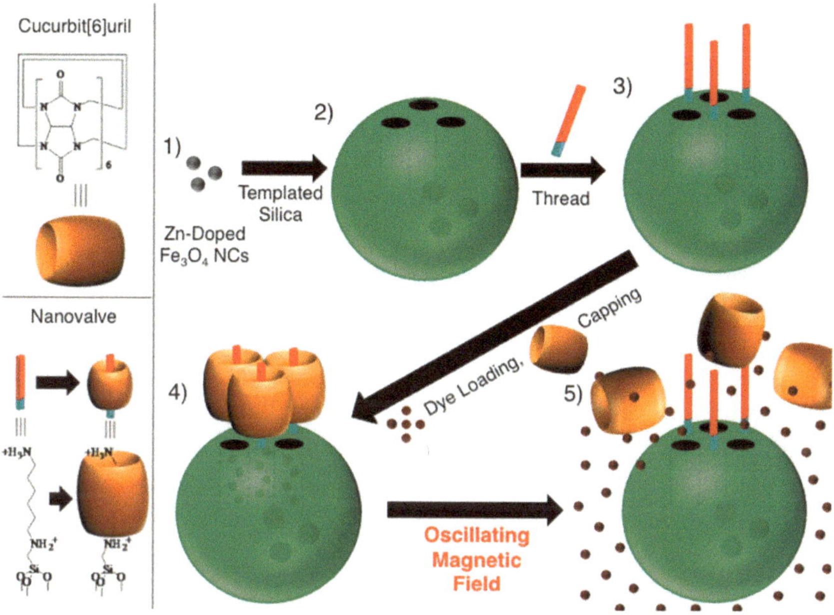

Fig. 9 Alternating magnetic field activation of a nanovalve. The binding constant of the bulky cucurbit(6)uril cap on the diamine stalk is temperature dependent. On exposure of the iron oxide-core MSN to the AMF the internal temperature increases causing release of the cap and the rhodamine B or doxorubicin cargo. *Adapted with permission from C.R. Thomas, D.P. Ferris, J.-H. Lee, E. Choi, M.H. Cho, E.S. Kim, J.F. Stoddart, J.-S. Shin, J. Cheon, J.I. Zink, Noninvasive remote-controlled release of drug molecules* in vitro *using magnetic actuation of mechanized nanoparticles, J. Am. Chem. Soc. 132 (2010) 10623–10625. Copyright (2010) American Chemical Society.*

An example of a small molecular nanovalve was reported by Rühle et al., who attached a thermolabile stalk and blocked the pores of SPION@MSN with a bulky gatekeeper forming by the supramolecular interactions between β-cyclodextrin and adamantane [107]. The heat generated by the AMF unblocked the pores by triggering a cycloreversion reaction (retro-Diels Alder reaction), dissociating the bulky gatekeepers from the silica surface and releasing the fluorescein molecules loaded in the pores.

3.2.3 DNA Dehybridization

Deoxyribonucleic acid (DNA) is responsive to many stimuli such as temperature, pH, enzymes, and competitive binding. Among them, the temperature trigger method is the most promising for clinical applications because it has spatial and temporal control of cargo release. When the temperature is above the DNA melting temperature, half of the DNA strands will be in the random coil or single-stranded state. Ruiz-Hernández et al. utilized this property to develop a nanovalve based on double-stranded DNA whose melting temperature is 47 °C [108]. Fluorescein molecules were entrapped in SPION@MSNs by complementary DNA. The authors showed that a reversible dehybridization/rehybridization of the DNA linker can be tuned by controlling the exposure to the AMF, leading to an "on–off" release mechanism of the cargo molecules. Instead of using SPION-functionalized DNA as a gatekeeper, Zhu et al. used DNA strands themselves to cap the pores of SPION@MSNs [109]. The authors investigated the synergistic effect of both the controlled release of DOX and hyperthermia. *In vitro* studies showed effective cell uptake of their particles in murine breast cancer 4T1 cells and negligible cytotoxicity.

3.2.4 Phase-Changing Polymers

Baeza et al. designed a thermo-sensitive block copolymer composed of poly(*N*-isopropylacrylamide) (PNIPAM) crosslinked with branched polyethyleneimine (PEI) to cap the pores of SPION@MSNs [110]. PNIPAM is a widely-studied phase-change polymer that can switch reversibly from an expanded hydrophilic phase to a hydrophobic phase with reduced volume when the temperature rises above its lower critical solution temperature (LCST) of approximately 32 °C in water [111]. As a result of the polymer shrinkage upon exposure to an AMF, pores were unblocked and fluorescein molecules were released after 6 h of magnetic field actuation. The authors further showed that the release of fluorescein triggered by magnetic fields (at 33 °C) was higher than that triggered by conventional heating at 45 °C. This result was explained by the finding that the temperature inside the SPION@MSNs was about 20 °C higher than the temperature in the bulk solution [104]. Recently, the same group used this system in *in vivo* studies and found the number of particles needed for injection per mice (182 μg Fe/tumor) in order to kill cancer cells was lower than other published results, suggesting that there is a pronounced synergistic effect between the drug release and hyperthermia [112].

Poly[(ethylene glycol)-*co*-(L-lactide)] (P(EO-*co*-LLA)) is another typical temperature sensitive polymer that exhibits reversible phase transitions between sol and gel as the temperature increases above its LCST [113]. Guo et al. grafted this thermal-sensitive polymer onto SPION@MSNs as a gatekeeper and demonstrated magnetic field-activated cargo release. They found that by combining the hyperthermia created by magnetic fields with the chemotherapy of an anticancer drug DOX, killing of HeLa cells was greatly enhanced compared to that of the free DOX and of the loaded particles without magnetic field activation [114].

4. FUTURE DIRECTIONS

Our review of stimuli-responsive release mechanisms emphasizes the nanomachines and caps that control the release, but they are not the only molecules that can be present. Many of the MSNs in this review carried not only the nanomachine, but also fluorescent dyes for imaging, targeting agents toward cancer cells, or all three. Future work in this area will undoubtedly use combinations of functionalities not only to enhance therapeutic applications but also for sensing leading to image guided therapy and for synergistic components to enhance the effect of drugs. As an example of useful combinations, a MSN based drug delivery system using pH-sensitive nanovalves and a transferrin (Tf) targeting agent improved drug delivery to cells with overexpressed transferrin receptors (TfR) [115]. An unusual permutation of components involved a low molecular weight polyethylene imine (PEI) and polyethylene imine-polyethylene glycol (PEI-PEG) copolymer, and pH-sensitive nanovalves to deliver siRNA. This serves as a promising method to achieve dual cargo delivery and controlled release [116].

An important goal will be to achieve specific recognition and response. An example in our review involved disease-specific antigen/antibody interactions [72]. Particles that not only release an imaging signal in the presence of a specified threat but also react to destroy it will be valuable. Particles that can "decide" whether or not to release a killing molecule could eliminate off-target damage. For example, a "logical" particle could recognize that it has been taken up into an acidic lysosome but not release a drug unless a second molecule specific to the disease was also present (using "AND" logic) [117].

An important advance would be to design mechanized particles that produce an amplified response when triggered. For example, a valve could open and expose a catalyst that produces large numbers of therapeutics using a design similar to the chemical amplifier [118]. The amplified response could also be used as a sensor to signal the presence of a threat (such as circulating cancer cells).

The next generation of stimuli-responsive particles may involve intercommunication between the particles. For example, a variety of "sentinel" particles, each specific for a threat, could individually detect the threat and send a general signal to an army of particles that could attack the threat.

Proof of concept for some of these applications has been published and the field is limited only by our imaginations. We expect in the future to see not only increasingly effective designs for drug delivery for specific diseases, but also synergistic combinations of materials with high efficacies.

ACKNOWLEDGMENTS

We thank the Defense Threat Reduction Agency Grant HDTRA1-13-1-0046 and the Zink Student Research Support fund for financial support.

REFERENCES

[1] M. Ebelmen, Sur la production artificielle de l'hydrophane, Comptes Rend. de I'Acd des Sciences 69 (1847) 527.
[2] C.J. Brinker, G.W. Scherer, Sol-Gel Science, Academic Press, New York, 1990.
[3] L. Ellerby, C. Nishida, F. Nishida, S. Yamanaka, B. Dunn, J. Valentine, J. Zink, Encapsulation of proteins in transparent porous silicate glasses prepared by the sol-gel method, Science 255 (1992) 1113–1115.
[4] K.E. Chung, E.H. Lan, M.S. Davidson, B.S. Dunn, J.S. Valentine, J.I. Zink, Measurement of dissolved oxygen in water using glass-encapsulated myoglobin, Anal. Chem. 67 (1995) 1505–1509.
[5] B.C. Dave, B. Dunn, J.S. Valentine, J.I. Zink, Sol-gel encapsulation methods for biosensors, Anal. Chem. 66 (2008) 1120A–1127A.
[6] J.S. Beck, C.T.-W. Chu, I.D. Johnson, C.T. Kresge, M.E. Leonowicz, W.J. Roth, J.W. Vartuli, Synthetic porous crystalline material: its synthesis and use WO Patent 91/1130, 1991.
[7] T. Yanagisawa, T. Shimizu, K. Kuroda, C. Kato, The preparation of alkyltriinethylaininonium-kaneinite complexes and their conversion to microporous materials, Bull. Chem. Soc. Jpn. 63 (1990) 988–992.
[8] V. Chiola, J.E. Ritsko, C.D. Vanderpool, Process for producing low bulk density silica, U.S. Patent 3, 1971, pp. 556–725.
[9] M.H. Huang, B.S. Dunn, J.I. Zink, In situ luminescence probing of the chemical and structural changes during formation of dip-coated lamellar phase sodium dodecyl sulfate sol – gel thin films, J. Am. Chem. Soc. 122 (2000) 3739–3745.
[10] F. Nishida, J.M. McKiernan, B. Dunn, J.I. Zink, *In situ* fluorescence probing of the chemical changes during sol-gel thin film formation, J. Am. Ceram. Soc. 78 (1995) 1640–1648.
[11] M.H. Huang, B.S. Dunn, H. Soyez, J.I. Zink, In situ probing by fluorescence spectroscopy of the formation of continuous highly-ordered lamellar-phase mesostructured thin films, Langmuir 14 (1998) 7331–7333.
[12] B.C. Dave, H. Soyez, J.M. Miller, B. Dunn, J.S. Valentine, J.I. Zink, Synthesis of protein-doped sol-gel SiO_2 thin films: evidence for rotational mobility of encapsulated cytochrome *c*, Chem. Mater. 7 (1995) 1431–1434.
[13] Y. Lu, R. Ganguli, C.A. Drewien, M.T. Anderson, C.J. Brinker, W. Gong, Y. Guo, H. Soyez, B. Dunn, M.H. Huang, J.I. Zink, Continuous formation of supported cubic and hexagonal mesoporous films by sol-gel dip-coating, Nature 389 (1997) 364–368.

[14] D. Tarn, C.E. Ashley, M. Xue, E.C. Carnes, J.I. Zink, C.J. Brinker, Mesoporous silica nanoparticle nano-carriers: biofunctionality and biocompatibility, Acc. Chem. Res. 46 (2013) 792–801.
[15] E. Aznar, M. Oroval, L. Pascual, J.R. Murguía, R. Martínez-Máñez, F. Sancenón, Gated materials for on-command release of guest molecules, Chem. Rev. 116 (2016) 561–718.
[16] B. Rühle, P. Saint-Cricq, J.I. Zink, Externally controlled nanomachines on mesoporous silica nanoparticles for biomedical applications, Chemphyschem 17 (2016) 1769–1779.
[17] F. Peng, Y. Su, Y. Zhong, C. Fan, S.-T. Lee, Y. He, Silicon nanomaterials platform for bioimaging, biosensing, and cancer therapy, Acc. Chem. Res. 47 (2014) 612–623.
[18] V. Biju, Chemical modifications and bioconjugate reactions of nanomaterials for sensing, imaging, drug delivery and therapy, Chem. Soc. Rev. 43 (2014) 744–764.
[19] Z. Tao, Mesoporous silica-based nanodevices for biological applications, RSC Adv. 4 (2014) 18961–18980.
[20] C. Argyo, V. Weiss, C. Bräuchle, T. Bein, Multifunctional mesoporous silica nanoparticles as a universal platform for drug delivery, Chem. Mater. 26 (2013) 435–451.
[21] Y. Chen, H. Chen, J. Shi, *In vivo* bio-safety evaluations and diagnostic/therapeutic applications of chemically designed mesoporous silica nanoparticles, Adv. Mater. 25 (2013) 3144–3176.
[22] J.L. Vivero-Escoto, R.C. Huxford-Phillips, W. Lin, Silica-based nanoprobes for biomedical imaging and theranostic applications, Chem. Soc. Rev. 41 (2012) 2673–2685.
[23] Z. Li, J.C. Barnes, A. Bosoy, J.F. Stoddart, J.I. Zink, Mesoporous silica nanoparticles in biomedical applications, Chem. Soc. Rev. 41 (2012) 2590–2605.
[24] J.E. Lee, N. Lee, T. Kim, J. Kim, T. Hyeon, Multifunctional mesoporous silica nanocomposite nanoparticles for theranostic applications, Acc. Chem. Res. 44 (2011) 893–902.
[25] M.W. Ambrogio, C.R. Thomas, Y.-L. Zhao, J.I. Zink, J.F. Stoddart, Mechanized silica nanoparticles: a new frontier in theranostic nanomedicine, Acc. Chem. Res. 44 (2011) 903–913.
[26] H. Meng, M. Xue, T. Xia, Y.-L. Zhao, F. Tamanoi, J.F. Stoddart, J.I. Zink, A.E. Nel, Autonomous *in vitro* anticancer drug release from mesoporous silica nanoparticles by pH-sensitive nanovalves, J. Am. Chem. Soc. 132 (2010) 12690–12697.
[27] Z. Li, D.L. Clemens, B.-Y. Lee, B.J. Dillon, M.A. Horwitz, J.I. Zink, Mesoporous silica nanoparticles with pH-sensitive nanovalves for delivery of moxifloxacin provide improved treatment of lethal pneumonic tularemia, ACS Nano 9 (2015) 10778–10789.
[28] D.L. Clemens, B.-Y. Lee, M. Xue, C.R. Thomas, H. Meng, D. Ferris, A.E. Nel, J.I. Zink, M.A. Horwitz, Targeted intracellular delivery of antituberculosis drugs to mycobacterium tuberculosis-infected macrophages via functionalized mesoporous silica nanoparticles, Antimicrob. Agents Chemother. 56 (2012) 2535–2545.
[29] S. Angelos, N.M. Khashab, Y.-W. Yang, A. Trabolsi, H.A. Khatib, J.F. Stoddart, J.I. Zink, pH Clock-operated mechanized nanoparticles, J. Am. Chem. Soc. 131 (2009) 12912–12914.
[30] D. Tarn, M. Xue, J.I. Zink, pH-Responsive dual cargo delivery from mesoporous silica nanoparticles with a metal-latched nanogate, Inorg. Chem. 52 (2013) 2044–2049.
[31] D. Tarn, C.-J. Yu, J. Lu, A. Hartz, F. Tamanoi, J.I. Zink, *In vitro* delivery of calcium ions by nanogated mesoporous silica nanoparticles to induce cancer cellular apoptosis, Mol. Sys. Des. Eng. 2 (2017) 384–392.

[32] S. Wu, Q. Deng, X. Huang, X. Du, Synergetic gating of metal-latching ligands and metal-chelating proteins for mesoporous silica nanovehicles to enhance delivery efficiency, ACS Appl. Mater. Interfaces 6 (2014) 15217–15223.
[33] C. Théron, A. Gallud, C. Carcel, M. Gary-Bobo, M. Maynadier, M. Garcia, J. Lu, F. Tamanoi, J.I. Zink, M. Wong Chi Man, Hybrid mesoporous silica nanoparticles with pH-operated and complementary H-bonding caps as an autonomous drug-delivery system, Chem. A Eur. J. 20 (2014) 9372–9380.
[34] C. Théron, A. Gallud, S. Giret, M. Maynadier, D. Grégoire, P. Puche, E. Jacquet, G. Pop, O. Sgarbura, V. Bellet, U. Hibner, J.I. Zink, M. Garcia, M. Wong Chi Man, C. Carcel, M. Gary-Bobo, pH-operated hybrid silica nanoparticles with multiple H-bond stoppers for colon cancer therapy, RSC Adv. 5 (2015) 64932–64936.
[35] Y.-L. Zhao, Z. Li, S. Kabehie, Y.Y. Botros, J.F. Stoddart, J.I. Zink, pH-Operated nanopistons on the surfaces of mesoporous silica nanoparticles, J. Am. Chem. Soc. 132 (2010) 13016–13025.
[36] M. Xue, D. Cao, J.F. Stoddart, J.I. Zink, Size-selective pH-operated megagates on mesoporous silica materials, Nanoscale 4 (2012) 7569–7574.
[37] C. Acosta, E. Pérez-Esteve, C.A. Fuenmayor, S. Benedetti, M.S. Cosio, J. Soto, F. Sancenón, S. Mannino, J. Barat, M.D. Marcos, R. Martínez-Máñez, Polymer composites containing gated mesoporous materials for on-command controlled release, ACS Appl. Mater. Interfaces 6 (2014) 6453–6460.
[38] Q. Gao, Y. Xu, D. Wu, W. Shen, F. Deng, Synthesis, characterization, and *in vitro* pH-controllable drug release from mesoporous silica spheres with switchable gates, Langmuir 26 (2010) 17133–17138.
[39] N.V. Roik, L.A. Belyakova, Chemical design of pH-sensitive nanovalves on the outer surface of mesoporous silicas for controlled storage and release of aromatic amino acid, J. Solid State Chem. 215 (2014) 284–291.
[40] V. Cauda, C. Argyo, A. Schlossbauer, T. Bein, Controlling the delivery kinetics from colloidal mesoporous silica nanoparticles with pH-sensitive gates, J. Mater. Chem. 20 (2010) 4305–4311.
[41] C. Wang, P. Lv, W. Wei, S. Tao, T. Hu, J. Yang, C. Meng, A smart multifunctional nanocomposite for intracellular targeted drug delivery and self-release, Nanotechnology 22 (2011) 415101.
[42] L. Xing, H. Zheng, Y. Cao, S. Che, Coordination polymer coated mesoporous silica nanoparticles for pH-responsive drug release, Adv. Mater. 24 (2012) 6433–6437.
[43] J. Wu, M.J. Sailor, Chitosan hydrogel-capped porous SiO_2 as a pH responsive nano-valve for triggered release of insulin, Adv. Funct. Mater. 19 (2009) 733–741.
[44] Z. Deng, Z. Zhen, X. Hu, S. Wu, Z. Xu, P.K. Chu, Hollow chitosan–silica nanospheres as pH-sensitive targeted delivery carriers in breast cancer therapy, Biomaterials 32 (2011) 4976–4986.
[45] X. Hu, Y. Wang, B. Peng, Chitosan-capped mesoporous silica nanoparticles as pH-responsive nano-carriers for controlled drug release, Chem. Asian J. 9 (2014) 319–327.
[46] J. Fu, Y. Zhu, Y. Zhao, Controlled free radical generation against tumor cells by pH-responsive mesoporous silica nanocomposite, J. Mater. Chem. B 2 (2014) 3538–3548.
[47] T. Xia, M. Kovochich, M. Liong, H. Meng, S. Kabehie, S. George, J.I. Zink, A.E. Nel, Polyethyleneimine coating enhances the cellular uptake of mesoporous silica nanoparticles and allows safe delivery of siRNA and DNA constructs, ACS Nano 3 (2009) 3273–3286.
[48] H. Meng, M. Liong, T. Xia, Z. Li, Z. Ji, J.I. Zink, A.E. Nel, Engineered design of mesoporous silica nanoparticles to deliver doxorubicin and P-Glycoprotein siRNA to overcome drug resistance in a cancer cell line, ACS Nano 4 (2010) 4539–4550.

[49] H. Meng, Y. Zhao, J. Dong, M. Xue, Y.-S. Lin, Z. Ji, W.X. Mai, H. Zhang, C.H. Chang, C.J. Brinker, J.I. Zink, A.E. Nel, Two-wave nanotherapy to target the stroma and optimize gemcitabine delivery to a human pancreatic cancer model in mice, ACS Nano 7 (2013) 10048–10065.
[50] H. Meng, W.X. Mai, H. Zhang, M. Xue, T. Xia, S. Lin, X. Wang, Y. Zhao, Z. Ji, J.I. Zink, A.E. Nel, Codelivery of an optimal drug/siRNA combination using mesoporous silica nanoparticles to overcome drug resistance in breast cancer *in vitro* and *in vivo*, ACS Nano 7 (2013) 994–1005.
[51] J.M. Rosenholm, E. Peuhu, J.E. Eriksson, C. Sahlgren, M. Lindén, Targeted intracellular delivery of hydrophobic agents using mesoporous hybrid silica nanoparticles as carrier systems, Nano Lett. 9 (2009) 3308–3311.
[52] X. Ma, Y. Zhao, K.W. Ng, Y. Zhao, Integrated hollow mesoporous silica nanoparticles for target drug/siRNA co-delivery, Chem. A Eur. J. 19 (2013) 15593–15603.
[53] Y. Zhu, J. Shi, W. Shen, X. Dong, J. Feng, M. Ruan, Y. Li, Stimuli-responsive controlled drug release from a hollow mesoporous silica sphere/polyelectrolyte multilayer core-shell structure, Angew. Chem. Int. Ed. 44 (2005) 5083–5087.
[54] Y.-J. Yang, X. Tao, Q. Hou, Y. Ma, X.-L. Chen, J.-F. Chen, Mesoporous silica nanotubes coated with multilayered polyelectrolytes for pH-controlled drug release, Acta Biomater. 6 (2010) 3092–3100.
[55] L. Minati, V. Antonini, M. Dalla Serra, G. Speranza, F. Enrichi, P. Riello, pH-activated doxorubicin release from polyelectrolyte complex layer coated mesoporous silica nanoparticles, Microporous Mesoporous Mater. 180 (2013) 86–91.
[56] S.Y. Chia, J.G. Cao, J.F. Stoddart, J.I. Zink, Working supramolecular machines trapped in glass and mounted on a film surface, Angew. Chem. Int. Ed. 40 (2001) 2447–2451.
[57] T.D. Nguyen, H.R. Tseng, P.C. Celestre, A.H. Flood, Y. Liu, J.F. Stoddart, J.I. Zink, A reversible molecular valve, Proc. Natl. Acad. Sci. 102 (2005) 10029–10034.
[58] T.D. Nguyen, Y. Liu, S. Saha, K.C.F. Leung, J.F. Stoddart, J.I. Zink, Design and optimization of molecular nanovalves based on redox-switchable bistable rotaxanes, J. Am. Chem. Soc. 129 (2007) 626–634.
[59] C.-Y. Lai, B.G. Trewyn, D.M. Jeftinija, K. Jeftinija, S. Xu, S. Jeftinija, V.S.Y. Lin, A mesoporous silica nanosphere-based carrier system with chemically removable CdS nanoparticle caps for stimuli-responsive controlled release of neurotransmitters and drug molecules, J. Am. Chem. Soc. 125 (2003) 4451–4459.
[60] M.W. Ambrogio, T.A. Pecorelli, K. Patel, N.M. Khashab, A. Trabolsi, H.A. Khatib, Y.Y. Botros, J.I. Zink, J.F. Stoddart, Snap-top nano-carriers, Org. Lett. 12 (2010) 3304–3307.
[61] B.-Y. Lee, Z. Li, D.L. Clemens, B.J. Dillon, A.A. Hwang, J.I. Zink, M.A. Horwitz, Redox-triggered release of moxifloxacin from mesoporous silica nanoparticles functionalized with disulfide snap-tops enhances efficacy against pneumonic tularemia in mice, Small 12 (2016) 3690–3702.
[62] F. Porta, G.E.M. Lamers, J.I. Zink, A. Kros, Peptide modified mesoporous silica nanocontainers, PCCP 13 (2011) 9982–9985.
[63] N.M. Khashab, A. Trabolsi, Y.A. Lau, M.W. Ambrogio, D.C. Friedman, H.A. Khatib, J.I. Zink, J.F. Stoddart, Redox- and pH-controlled mechanized nanoparticles, Eur. J. Org. Chem. 2009 (2009) 1669–1673.
[64] P. Shi, Y. Qu, C. Liu, H. Khan, P. Sun, W. Zhang, Redox-responsive multicompartment vesicles of ferrocene-containing triblock terpolymer exhibiting on–off switchable pores, ACS Macro Lett. 5 (2015) 88–93.

[65] K. Patel, S. Angelos, W.R. Dichtel, A. Coskun, Y.-W. Yang, J.I. Zink, J.F. Stoddart, Enzyme-responsive snap-top covered silica nanocontainers, J. Am. Chem. Soc. 130 (2008) 2382–2383.
[66] Y.-J. Cheng, G.-F. Luo, J.-Y. Zhu, X.-D. Xu, X. Zeng, D.-B. Cheng, Y.-M. Li, Y. Wu, X.-Z. Zhang, R.-X. Zhuo, F. He, Enzyme-induced and tumor-targeted drug delivery system based on multifunctional mesoporous silica nanoparticles, ACS Appl. Mater. Interfaces 7 (2015) 9078–9087.
[67] B. Kumar, S. Kulanthaivel, A. Mondal, S. Mishra, B. Banerjee, A. Bhaumik, I. Banerjee, S. Giri, Mesoporous silica nanoparticle based enzyme responsive system for colon specific drug delivery through guar gum capping, Colloids Surf. B Biointerfaces 150 (2017) 352–361.
[68] K. Radhakrishnan, S. Gupta, D.P. Gnanadhas, P.C. Ramamurthy, D. Chakravortty, A.M. Raichur, Protamine-capped mesoporous silica nanoparticles for biologically triggered drug release, Part. Part. Syst. Charact. 31 (2014) 449–458.
[69] L. Mondragón, N. Mas, V. Ferragud, C. de la Torre, A. Agostini, R. Martínez-Máñez, F. Sancenón, P. Amorós, E. Pérez-Payá, M. Orzáez, Enzyme-responsive intracellular-controlled release using silica mesoporous nanoparticles capped with ε-poly-L-lysine, Chem. A Eur. J. 20 (2014) 5271–5281.
[70] J.L. Vivero-Escoto, L. Moore Jeffords, D. Dréau, M. Alvarez-Berrios, P. Mukherjee, Mucin1 antibody-conjugated dye-doped mesoporous silica nanoparticles for breast cancer detection *in vivo*, Proc. SPIE 10078 (2017) 100780B1–100780B8.
[71] S. Zhou, D. Wu, X. Yin, X. Jin, X. Zhang, S. Zheng, C. Wang, Y. Liu, Intracellular - pH-responsive and rituximab-conjugated mesoporous silica nanoparticles for targeted drug delivery to lymphoma B cells, J. Exp. Clin. Cancer Res. 36 (2017) 24.
[72] B. Ruehle, D.L. Clemens, B.-Y. Lee, M.A. Horwitz, J.I. Zink, A pathogen-specific cargo delivery platform based on mesoporous silica nanoparticles, J. Am. Chem. Soc. 139 (2017) 6663–6668.
[73] R. Bhat, I. García, E. Aznar, B. Arnaiz, M.C. Martínez-Bisbal, L.M. Liz-Marzán, S. Penadés, R. Martínez-Máñez, Lectin-gated and glycan functionalized mesoporous silica nanocontainers for targeting cancer cells overexpressing Lewis X antigen, Nanoscale 10 (2018) 239–249.
[74] W. Gao, J.-x. Liang, Y. Liang, Clinicopathological and prognostic significance of sialyl Lewis X overexpression in patients with cancer: a meta-analysis, Onco Targets Ther. 9 (2016) 3113–3125.
[75] P. Sozzani, R. Arisio, M. Porpiglia, C. Benedetto, Is Sialyl Lewis x antigen expression a prognostic factor in patients with breast cancer? Int. J. Surg. Pathol. 16 (2008) 365–374.
[76] K. Nagao, Y. Itoh, K. Fujita, M. Fujime, Evaluation of urinary CA19-9 levels in bladder cancer patients classified according to the combinations of Lewis and Secretor blood group genotypes, Int. J. Urol. 14 (2007) 795–799.
[77] P. Sierocki, H. Maas, P. Dragut, G. Richardt, F. Vögtle, L. De Cola, F. Brouwer, J.I. Zink, Photoisomerization of azobenzene derivatives in nanostructured silica, J. Phys. Chem. B. 110 (2006) 24390–24398.
[78] Y. Zhu, M. Fujiwara, Installing dynamic molecular photomechanics in mesopores: a multifunctional controlled-release nanosystem, Angew. Chem. Int. Ed. 46 (2007) 2241–2244.
[79] S. Angelos, E. Choi, F. Vögtle, L. De Cola, J.I. Zink, Photo-driven expulsion of molecules from mesostructured silica nanoparticles, J. Phys. Chem. C 111 (2007) 6589–6592.
[80] J. Lu, E. Choi, F. Tamanoi, J.I. Zink, Light-activated nanoimpeller-controlled drug release in cancer cells, Small 4 (2008) 421–426.

[81] Y.A. Lau, B.L. Henderson, J. Lu, D.P. Ferris, F. Tamanoi, J.I. Zink, Continuous spectroscopic measurements of photo-stimulated release of molecules by nanomachines in a single living cell, Nanoscale 4 (2012) 3482–3489.
[82] D.P. Ferris, Y.-L. Zhao, N.M. Khashab, H.A. Khatib, J.F. Stoddart, J.I. Zink, Light-operated mechanized nanoparticles, J. Am. Chem. Soc. 131 (2009) 1686–1688.
[83] D. Tarn, D.P. Ferris, J.C. Barnes, M.W. Ambrogio, J.F. Stoddart, J.I. Zink, A reversible light-operated nanovalve on mesoporous silica nanoparticles, Nanoscale 6 (2014) 3335–3343.
[84] D. Wang, S. Wu, Red-light-responsive supramolecular valves for photocontrolled drug release from mesoporous nanoparticles, Langmuir 32 (2016) 632–636.
[85] Q. Yuan, Y. Zhang, T. Chen, D. Lu, Z. Zhao, X. Zhang, Z. Li, C.-H. Yan, W. Tan, Photon-manipulated drug release from a mesoporous nanocontainer controlled by azobenzene-modified nucleic acid, ACS Nano 6 (2012) 6337–6344.
[86] C. Hu, K.R. West, O.A. Scherman, Hollow mesoporous raspberry-like colloids with removable caps as photoresponsive nanocontainers, Nanoscale 8 (2016) 7840–7844.
[87] A. Agostini, F. Sancenón, R. Martínez-Máñez, M.D. Marcos, J. Soto, P. Amorós, A photoactivated molecular gate, Chem. A Eur. J. 18 (2012) 12218–12221.
[88] T.M. Guardado-Alvarez, L. Sudha Devi, M.M. Russell, B.J. Schwartz, J.I. Zink, Activation of snap-top capped mesoporous silica nanocontainers using two near-infrared photons, J. Am. Chem. Soc. 135 (2013) 14000–14003.
[89] J. Lai, X. Mu, Y. Xu, X. Wu, C. Wu, C. Li, J. Chen, Y. Zhao, Light-responsive nanogated ensemble based on polymer grafted mesoporous silica hybrid nanoparticles, Chem. Commun. 46 (2010) 7370–7372.
[90] J. Croissant, J.I. Zink, Nanovalve-controlled cargo release activated by plasmonic heating, J. Am. Chem. Soc. 134 (2012) 7628–7631.
[91] H. Li, L.-L. Tan, P. Jia, Q.-L. Li, Y.-L. Sun, J. Zhang, Y.-Q. Ning, J. Yu, Y.-W. Yang, Near-infrared light-responsive supramolecular nanovalve based on mesoporous silica-coated gold nanorods, Chem. Sci. 5 (2014) 2804–2808.
[92] J. Liu, C. Detrembleur, M.-C. De Pauw-Gillet, S. Mornet, C. Jérôme, E. Duguet, Gold nanorods coated with mesoporous silica shell as drug delivery system for remote near infrared light-activated release and potential phototherapy, Small 11 (2015) 2323–2332.
[93] Z. Zhang, L. Wang, J. Wang, X. Jiang, X. Li, Z. Hu, Y. Ji, X. Wu, C. Chen, Mesoporous silica-coated gold nanorods as a light-mediated multifunctional theranostic platform for cancer treatment, Adv. Mater. 24 (2012) 1418–1423.
[94] Z. Song, J. Shi, Z. Zhang, Z. Qi, S. Han, S. Cao, Mesoporous silica-coated gold nanorods with a thermally responsive polymeric cap for near-infrared-activated drug delivery, J. Mater. Sci. 53 (2018) 7165–7179.
[95] J.G. Croissant, C. Qi, O. Mongin, V. Hugues, M. Blanchard-Desce, L. Raehm, X. Cattoën, M. Wong Chi Man, M. Maynadier, M. Gary-Bobo, M. Garcia, J.I. Zink, J.-O. Durand, Disulfide-gated mesoporous silica nanoparticles designed for two-photon-triggered drug release and imaging, J. Mater. Chem. B 3 (2015) 6456–6461.
[96] W.H. Thompson, Solvation dynamics and proton transfer in nanoconfined liquids, Annu. Rev. Phys. Chem. 62 (2011) 599–619.
[97] T.M. Guardado-Alvarez, M.M. Russell, J.I. Zink, Nanovalve activation by surface-attached photoacids, Chem. Commun. 50 (2014) 8388–8390.
[98] A. Noureddine, L. Lichon, M. Maynadier, M. Garcia, M. Gary-Bobo, J.I. Zink, X. Cattoën, M. Wong Chi Man, Controlled multiple functionalization of mesoporous silica nanoparticles: homogeneous implementation of pairs of functionalities communicating through energy or proton transfers, Nanoscale 7 (2015) 11444–11452.

[99] T.D. Nguyen, K.C.F. Leung, M. Liong, Y. Liu, J.F. Stoddart, J.I. Zink, Versatile supramolecular nanovalves reconfigured for light activation, Adv. Funct. Mater. 17 (2007) 2101–2110.

[100] T.M. Guardado-Alvarez, L.S. Devi, J.-M. Vabre, T.A. Pecorelli, B.J. Schwartz, J.-O. - Durand, O. Mongin, M. Blanchard-Desce, J.I. Zink, Photo-redox activated drug delivery systems operating under two photon excitation in the near-IR, Nanoscale 6 (2014) 4652–4658.

[101] S. Saha, E. Johansson, A.H. Flood, H.-R. Tseng, J.I. Zink, J.F. Stoddart, A photoactive molecular triad as a nanoscale power supply for a supramolecular machine, Chem. A Eur. J. 11 (2005) 6846–6858.

[102] J. Croissant, M. Maynadier, A. Gallud, H. Peindy N'Dongo, J.L. Nyalosaso, G. Derrien, C. Charnay, J.-O. Durand, L. Raehm, F. Serein-Spirau, N. Cheminet, T. Jarrosson, O. Mongin, M. Blanchard-Desce, M. Gary-Bobo, M. Garcia, J. Lu, F. Tamanoi, D. Tarn, T.M. Guardado-Alvarez, J.I. Zink, Two-photon-triggered drug delivery in cancer cells using nanoimpellers, Angew. Chem. Int. Ed. 125 (2013) 14058–14062.

[103] S. Laurent, S. Dutz, U.O. Häfeli, M. Mahmoudi, Magnetic fluid hyperthermia: Focus on superparamagnetic iron oxide nanoparticles, Adv. Colloid Interface Sci. 166 (2011) 8–23.

[104] J. Dong, J.I. Zink, Taking the temperature of the interiors of magnetically heated nanoparticles, ACS Nano 8 (2014) 5199–5207.

[105] C.R. Thomas, D.P. Ferris, J.-H. Lee, E. Choi, M.H. Cho, E.S. Kim, J.F. Stoddart, J.-S. Shin, J. Cheon, J.I. Zink, Noninvasive remote-controlled release of drug molecules *in vitro* using magnetic actuation of mechanized nanoparticles, J. Am. Chem. Soc. 132 (2010) 10623–10625.

[106] P. Saint-Cricq, S. Deshayes, J.I. Zink, A.M. Kasko, Magnetic field activated drug delivery using thermodegradable azo-functionalised PEG-coated core–shell mesoporous silica nanoparticles, Nanoscale 7 (2015) 13168–13172.

[107] B. Rühle, S. Datz, C. Argyo, T. Bein, J.I. Zink, A molecular nanocap activated by superparamagnetic heating for externally stimulated cargo release, Chem. Commun. 52 (2016) 1843–1846.

[108] E. Ruiz-Hernández, A. Baeza, M. Vallet-Regí, Smart drug delivery through DNA/magnetic nanoparticle gates, ACS Nano 5 (2011) 1259–1266.

[109] Y. Zhu, C. Tao, DNA-capped Fe3O4/SiO2 magnetic mesoporous silica nanoparticles for potential controlled drug release and hyperthermia, RSC Adv. 5 (2015) 22365–22372.

[110] A. Baeza, E. Guisasola, E. Ruiz-Hernández, M. Vallet-Regí, Magnetically triggered multidrug release by hybrid mesoporous silica nanoparticles, Chem. Mater. 24 (2012) 517–524.

[111] C. Liu, J. Guo, W. Yang, J. Hu, C. Wang, S. Fu, Magnetic mesoporous silica microspheres with thermo-sensitive polymer shell for controlled drug release, J. Mater. Chem. 19 (2009) 4764–4770.

[112] E. Guisasola, L. Asín, L. Beola, J.M. de la Fuente, A. Baeza, M. Vallet-Regí, Beyond traditional hyperthermia: *in vivo* cancer treatment with magnetic-responsive mesoporous silica nano-carriers, ACS Appl. Mater. Interfaces 10 (2018) 12518–12525.

[113] B. Jeong, Y.H. Bae, D.S. Lee, S.W. Kim, Biodegradable block copolymers as injectable drug-delivery systems, Nature 388 (1997) 860–862.

[114] W. Guo, C. Yang, H. Lin, F. Qu, P(EO-co-LLA) functionalized Fe_3O_4@mSiO_2 nanocomposites for thermo/pH responsive drug controlled release and hyperthermia, Dalton Trans. 43 (2014) 18056–18065.

[115] A.A. Hwang, J. Lu, F. Tamanoi, J.I. Zink, Functional nanovalves on protein-coated nanoparticles for *in vitro* and *in vivo* controlled drug delivery, Small 11 (2015) 319–328.

[116] J. Dong, M. Xue, J.I. Zink, Functioning of nanovalves on polymer coated mesoporous silica nanoparticles, Nanoscale 5 (2013) 10300–10306.
[117] S. Angelos, Y.-W. Yang, N.M. Khashab, J.F. Stoddart, J.I. Zink, Dual-controlled nanoparticles exhibiting AND logic, J. Am. Chem. Soc. 131 (2009) 11344–11346.
[118] M. Xue, J.I. Zink, An enzymatic chemical amplifier based on mechanized nanoparticles, J. Am. Chem. Soc. 135 (2013) 17659–17662.

CHAPTER FOUR

Mesoporous Silica-Based Nanoparticles for Light-Actuated Biomedical Applications via Near-Infrared Two-Photon Absorption

Jonas G. Croissant*,†,1, Jean-Olivier Durand‡
*Chemical and Biological Engineering, University of New Mexico, Albuquerque, NM, United States
†Center for Micro-Engineered Materials, Advanced Materials Laboratory, University of New Mexico, Albuquerque, NM, United States
‡Institut Charles Gerhardt Montpellier, UMR-5253 CNRS-UM-ENSCM, Montpellier, France
[1]Corresponding author: e-mail address: jonascroissant@unm.edu

Contents

The Enzymes, Volume 43
ISSN 1874-6047
https://doi.org/10.1016/bs.enz.2018.07.004

Abstract

In this review, we highlight the design of nanomaterials for two-photon excitation, in order to treat tumors with a high accuracy. Indeed two-photon excitation allows remote control of the nanoparticles with a spatio-temporal resolution. The nanomaterials are based on mesoporous silica-organosilica nanoparticles including core-shell systems. The therapeutic treatments include drug delivery, photodynamic therapy, gene silencing, and their combinations. At first, the nanosystems designed for two-photon-triggered cytotoxic drug delivery are reviewed. Then the nanomaterials prepared for two-photon photodynamic therapy and reactive oxygen species delivery are discussed. Finally, the nanosystems combining drug delivery or gene silencing with two-photon photodynamic therapy are presented. Due to the rapid progresses concerning two-photon-excited nanomaterials and the interest of near-infrared light to treat deep tumors, we believe this technology could be of high interest for the personalized medicine of the future.

1. INTRODUCTION

1.1 Benefits of Near-Infrared Absorption

Light-responsive nanoparticles have been increasingly studied for academic [1–7] and clinical biomedical research [8,9]. Light stimuli of various photophysical properties are indeed ideal triggers for drug delivery [3,4,10–17], bioimaging [14,18–20], photodynamic therapy [9,21–28], photothermal therapy [29–33], biosensing [34–36], and biocatalysis applications [10,37,38]. The advantages of light actuation, in general, include a remote control of the target bioresponse and a spatiotemporal resolution of the target site for selective effects [13,39]. Tuning the light parameters such as the wavelength, power, and duration can also be used to actuate specific absorbers as well as the intensity of the response [40,41].

Near-infrared (NIR) light absorption is necessary for effective biomedical applications since it minimizes the scattering and attenuation of the irradiation in the living tissues [42,43]. The graphic displayed in Fig. 1 plots the irradiation penetration depth in the human skin as a function of the wavelength. The penetration of the laser beam is highest between approximately 700 and 1300 nm which is why this region has been called the biological window [44–47]. The variation of the laser penetration is related to the different refractive indices and molar extinction coefficients of the components present in human tissues including the skin layer, the blood,

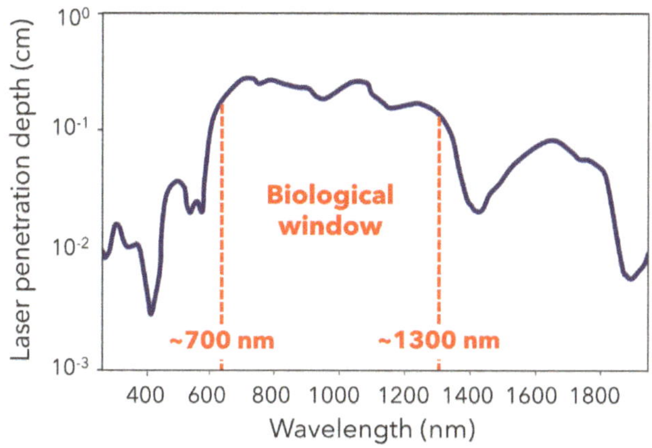

Fig. 1 Light penetration depth in the human skin as a function of the wavelength of the irradiation. The 700–1300 nm region has the highest penetration depth and is called the biological window.

water, and fat tissues [44]. The light scattering coefficient of human tissue also decreases as the wavelength increases, which accounts for the interests of NIR-actuated therapy. A depth of penetration of about 2 cm is obtained at a wavelength of 740 nm [48], which is thus well suited for a safe laser-actuated nanoparticle-based therapeutic and/or diagnostic effect [49].

1.2 Benefits of Two-Photon Absorption

Two-photon absorption is a rare phenomenon of nonlinear optics in which a molecule absorbs two photons simultaneously. In a two-photon absorption (TPA) process, two photons of identical or different frequencies are absorbed to excite a molecule from one state (typically the ground state) to a higher energy electronic state, and the energetic transition equals the sum of the energy of the two photons. Fig. 2 depicts the Jablonski diagrams of the one-photon and two-photon absorption processes in the case of TPA with two photons of equal energies in the case of TPA [50]. TPA differs from linear absorptions since its effectiveness varies with the square of the light intensity, whereas one-photon absorption varies linearly with the light intensity.

The TPA phenomenon was first predicted by Maria Goeppert-Mayer in 1931 in her doctoral dissertation [1]. It took the invention of the laser, 30 years later, for the experimental verification of the TPA using a

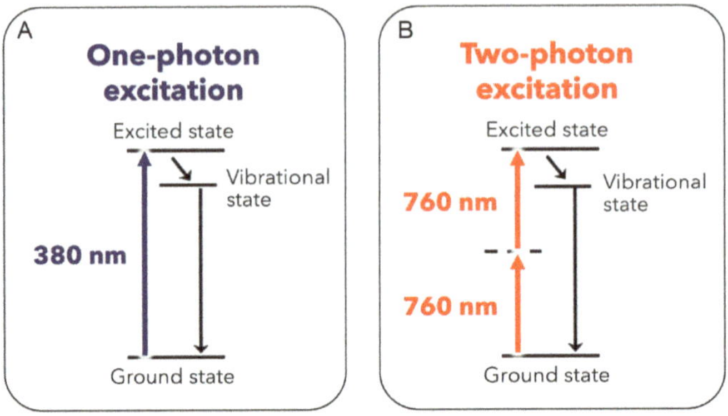

Fig. 2 Jablonski diagram of one (A) and two-photon excitations (B). Fluorescent absorption and emission transitions between the ground and excited states are depicted by vertical arrows.

europium-doped crystal [51,52]. Since the probability of the simultaneous absorption of two photons is extremely low (thus the absence of commercially available and effective absorbers), enhancing the TPA is paramount. TPE enhancement is obtained using high-power continuous wave lasers or short-pulse of femtosecond lasers [53–56]. In addition, confocal two-photon microscopy densifies photons in space and in time which further increases the likelihood of a TPA event.

The benefits of two-photon-actuated medical applications of nanotechnology are many. Considering the case of cancer therapy [1], for instance, TPA enables a three-dimensional resolution of the therapeutic effect (down to a micron cube) along with the time resolution associated with the duration of the irradiation. The tumor could, therefore, be screened by the laser beam with the precision toward a single cell which is especially important for many cancers such as brain and eye cancers. The non-fulfillment of these criteria seriously diminished the effectiveness of the therapy and produces unpredictable and unacceptable side effects on patients. Fig. 3 illustrates the low scattering and 3-D resolution of two-photon versus on-photon absorptions. In summary, two-photon-excited nanomedicine in the NIR is particularly relevant for cancer therapy and imaging [23,57–65]. Pathologies involving short laser penetrations such as skin and retinoblastoma cancers could greatly benefit from efficient medical nanodevices using TPA in the NIR. Increasing the laser depth penetration threshold, surgery, and the well-developed endoscopy technologies could be harnessed.

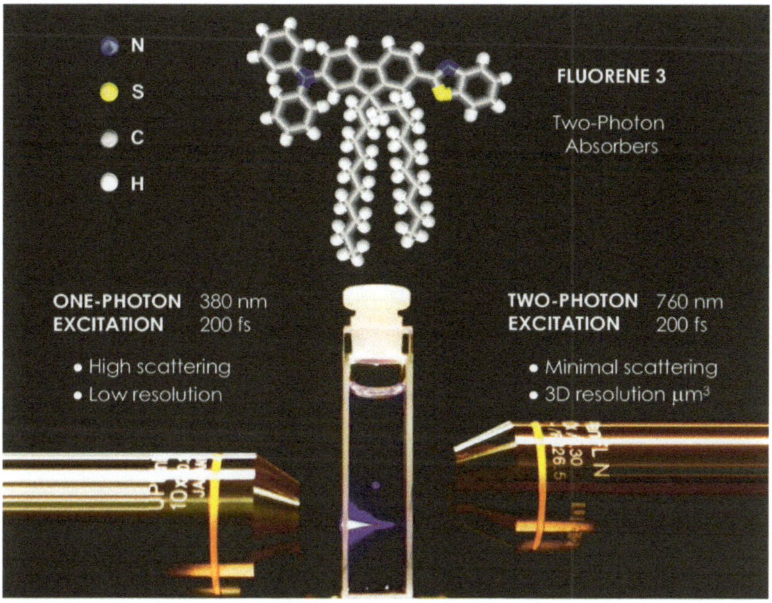

Fig. 3 Comparison of the one-photon and two-photon absorptions of fluorene 3 dyes upon pulsed laser excitations (200 fs) at 380 nm (left) and 760 nm (right). *Adapted photograph, courtesy of the group of Prof. K. D. Belfield.*

2. CONFIGURING NANOPARTICLES FOR TWO-PHOTON-ACTUATED THERAPY

Given all the benefits of the two-photon excitation in the NIR, how can NPs be configured for two-photon-triggered therapy such as drug delivery and photodynamic therapy? That is not a straightforward task since only a few studies succeeded to configure two-photon-triggered drug delivery [7,39,66–72].

2.1 Materials Design Strategies for Drug Delivery

The challenge of configuring drug delivery from silica nanomaterials [10,18,73–83] upon two-photon excitation is five-fold: (i) The organic synthesis challenge: preparing two-photon absorbers with high two-photon cross section [84], (ii) The nanoparticle-two-photon absorber challenge: preparing nanoparticles composed of two-photon absorbers while maintaining high σ_2 per chromophore [85], (iii) The two-photon-triggering of the cargo release: preparing cargo-loaded nanocarriers with a delivery mechanism using a two-photon-triggered physical-chemical effect, (iv) Monitoring the two-photon-triggered release: since the quantity released might be relatively limited, (v) Realizing the two-photon-triggered delivery in cells.

First, in order to address the first challenge, synthetic chemists have had to produce effective two-photon absorbers having σ_2 ranging from a hundred to several thousand GM since most molecules have extremely low σ_2. In the absence of this expertise, research groups have used commercially available coumarin precursors ($\sigma_2 \sim 1.07$ GM). The research of Anderson and co-workers has summarized the key ingredient for the design of effective two-photon absorbers: *"(1) Long, π-conjugated chains with enforced coplanarity [...],(2) Donor and acceptor groups at the center and ends of the molecule [...], (3) Centrosymmetric chromophores that possess a strong one-photon absorption transition close to the two-photon absorption laser wavelength [...], (4) Chromophores with narrow one- and two-photon absorption bands."* [86]

Secondly, fabricating nanoparticles composed of two-photon absorbers while maintaining high σ_2 per chromophore is also challenging since as we showed in one study [85], the resulting properties of the encapsulated two-photon chromophores are highly dependent on the synthetic methods, used as they correlate the spatial proximities of the absorbers. In addition, many organic absorbers are sensitive to acidic or basic conditions, to the temperature and to other synthesis-generated modifications, and their degradation hampers their photophysical properties.

Third, triggering of the cargo release upon TPA has been achieved by four main approaches, as summarized in Fig. 4. The first strategy consists in transferring energy from a nanoparticle core capable of TPA such as spatially close gold nanospheres, or gold nanorods. The transferred energy can then actuate various surface polymers or molecular gates [13]. The second approach is based on two-photon absorber-doped polymers coated onto silica particles. The polymer thus initially prevents the diffusion of the loaded cargos and is then photo-modified to trigger the release. The polymer change can be a photolytic process [15] leading to a polymeric disassembly. Two-photon-actuated photodissasembly without bond cleavage was also reported upon the photorearrangement of Wolff [66,87], which involves the transformation of hydrophobic diazonaphthoquinone into a hydrophilic derivative, which was used for the degradation of polymeric nanoassemblies to delivery drug molecules. The third approach uses molecular or supramolecular nanomachines engineered onto silica particles to block their drug-containing mesopores [10], and in combination with surface-grafted two-photon absorbers can be opened on demand by electron transfer

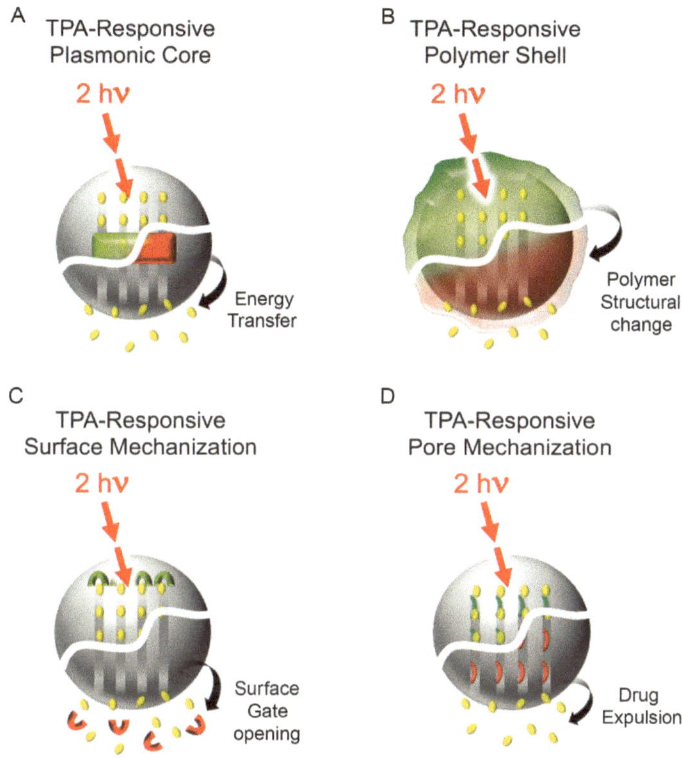

Fig. 4 Main materials design strategies for drug delivery via silica particles using two-photon responsive nanoparticle cores (A), polymer shells (B), surface molecular or supramolecular mechanizations (C), or pore mechanization (D).

mediated bond cleavage, direct bond cleavage, or energy transfer mechanisms. The fourth strategy mechanizes the pore interior to trap and releases drugs covalently (direct photolysis of pro-drug systems) or physically (e.g., drug expulsion upon photoisomerization) loaded into the nanoporous framework of silica hybrids.

Briefly comparing these approaches, it can be noted that they hold different advantages. For instance, using gold nanorod core@mesoporous silica shell nanocarriers, the gold plasmonic cores can be used for both bioimaging and photothermal therapy in addition to triggering the delivery of loaded drugs. The polymer coating is often a single-step coating step and can be most likely less sensitive than nanomachine syntheses against off-target biodegradations. However, it has been recently shown that nanomachine can still function in vivo after being coated with polymers [88]. The pro-drug design would theoretically have lowest passive leakage since the drugs are covalently linked inside the pores, but it is limited to specific cargos. Interestingly, the above-mentioned advantages can be combined in a single nanocarrier medical device according to the job description of the specific pharmaceutical-disease requirements.

2.2 Materials Design Strategies for Photodynamic Therapy

Similarly to the strategies for two-photon-mediated drug delivery, harnessing the TPA of photosensitizers can be performed by designing organic molecules with high σ_2 and incorporating them into nanoparticles. The effective biomedical application of photosensitizers is typically limited by their low water solubility, cellular membrane impermeability, and rapid biodegradation or biocapture, which is why photosensitizers require nanovectorization. To date, three approaches have been reported to vectorize molecular photosensitizers via nanoparticles: namely direct photosensitization of two-photon-absorbing photosensitizers (Fig. 5A) [89–95] and indirect photosensitization

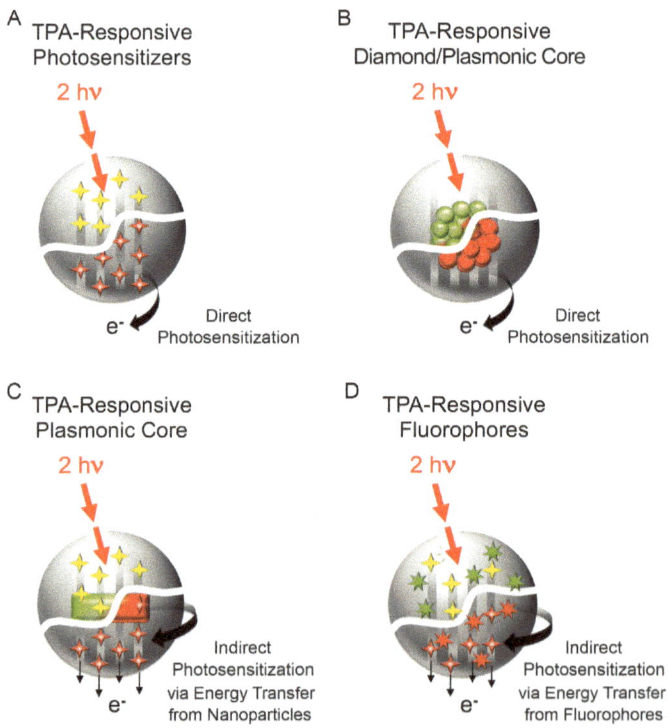

Fig. 5 Main materials design strategies for photodynamic therapy via silica particles using direct photosensitization using photosensitizers (A) or nanodiamond cores (B); or indirect photosensitization using plasmonic energy transfer (C) or energy transfer from chromophores (green to red stars) (D).

of effective one-photon-absorbing photosensitizers via an energy transfer through a plasmonic nanocore (Fig. 5C) [90,91] or chromophores with a large TPA σ_2 (Fig. 5D) [1]. Alternatively, direct electron production for photodynamic therapy has also been reported using nanodiamond or gold nanoparticle (σ_2^{MAX} ranging 4.10^3–1.10^4 GM) [1] cores as electron sources (Fig. 5B) [96].

3. TWO-PHOTON-EXCITED DRUG DELIVERY

3.1 Silica-Coumarin Nanocomposites

The group of Zhu first configured silica nanoparticles for TPA and two-photon-triggered drug delivery using organically doped mesoporous silica nanoparticles [72]. Their materials design involved a prod-drug mechanism with amino-functionalized MSN covalently binding to chlorambucil anti-cancer drug. Coumarin is commercially available and has a low $\sigma_2^{740\,nm}$ of 1.07 GM which provides a low and sufficient two-photon sensitivity for the proof of principle of delivery molecules upon TPA [1]. This design allowed the photo-induced drug release by cleavage of the carbonyl coumarin bond, as monitored in solution under pulsed laser excitation at 800 nm, and subsequently HeLa cancer cells under visible light exposure. This excellent work has a remarkable impact on the field and inspired more studies on non-drug specific two-photon-excited delivery using silica-based particles.

In order to overcome the limitations of pro-drug nanocarriers, the group of Zink investigated coumarin-based nanocaps engineered onto the surface of 100 nm-wide mesoporous silica nanoparticles [68]. Coumarin photoabsorbers were installed at the end of the molecular stalk and used to supramolecularly bind β-cyclodextrin rings which capped the pores of rhodamine B-loaded silica particles. The photolysis of coumarin bond was then triggered upon an 800 nm pulsed laser excitation (40 fs, 1 kHz, 60 μJ) which was followed by the cargo release from mesoporous silica nanoparticles [68].

Lu and co-workers then reported doxorubicin anticancer drug-loaded hollow mesoporous silica nanoparticles coated with coumarin-containing block copolymer (see Fig. 6) [15]. The amphiphilic copolymers were self-assembled onto the 250 nm-wide silica particles after drug loading for photocleavage-mediated delivery upon TPA of coumarin groups (Fig. 6A and B). The absence of passive leakage before the laser exposure showed

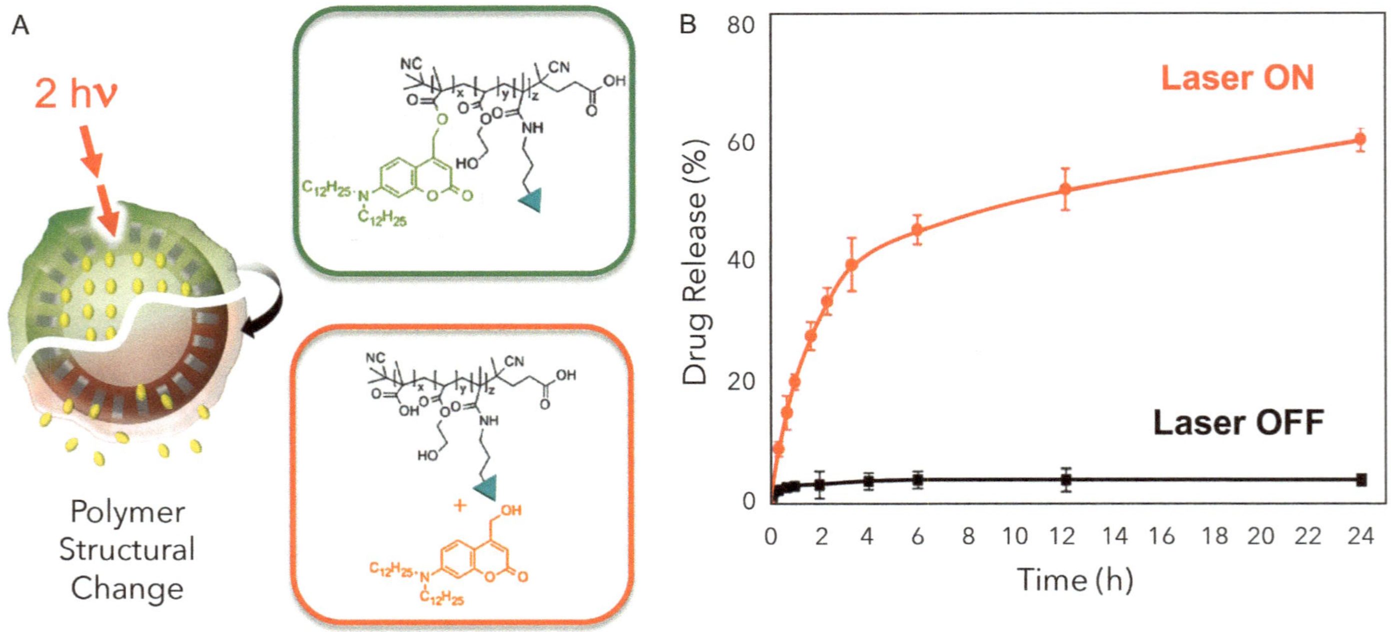

Fig. 6 Representation of the materials design of polymer-coumarin core-hollow silica shell nanoparticles for TPA-actuated drug delivery by degradation upon laser exposure (A). Drug release profile of particles with and without laser irradiation (B). *Illustration based on the finding of W. Ji, N. Li, D. Chen, X. Qi, W. Sha, Y. Jiao, Q. Xu, J. Lu, Coumarin-containing photo-responsive nanocomposites for NIR light-triggered controlled drug release via a two-photon process, J. Mater. Chem. B 1 (2013) 5942–5949.*

the tight protection of drugs into silica nanoparticles (Fig. 6B). The study showed 65% of KB cell death, although it required very high NPs concentration (1 mg/mL) and irradiation time (980 nm, 60 min). This was due to the low cross section of coumarin moieties in the copolymer, which displayed a time-dependent photolysis efficiency (ca. 70% after 100 min of light exposure at 800 nm, 1 kHz, 2.28 mJ/pulse).

3.2 Silica-Diphenyl-Butadiene Nanocomposites

In order to increase the sensibility of the TPA, restricted by the low two-photon cross-section of commercially available coumarin (~1.07 GM) and fluorescein absorbers (~38 GM), novel two-photon absorbers were prepared by multi-step organic synthesis. Durand and co-workers synthesized a tetra-alkoxysilylated di(aminophenyl)butadiene photosensitizer having a $\sigma_2^{740\ nm}$ of 110 GM [67]. The two-photon absorber was then co-condensed with tetraethoxysilane to obtain organosilica hybrid mesoporous nanocarriers containing two-photon-responsive electron generators inside the silica mesopore walls. The hybrid particles were designed with two-photon photosensitizer contents ranging from 7 to nearly 40 wt% to optimize the resulting photophysical properties of the nanocarriers [85]. The properties were found to depend heavily on the synthetic method as particles with identical photosensitizer contents obtained via different sol-gel processes [97–100]. Emission spectra of the nanoparticles demonstrated that this fact arose from spatial dispersion variation leading to photoquenching for the highest doping content reported. Best photophysical properties were obtained with a ~16 wt% doping leading to a higher $\sigma_2^{700\ nm}$ in the particles (200 GM) than in the precursor (110 GM), with a fluorescence quantum yield (Φ_F) of 41% and a brightness ($\sigma_2^{MAX} \times \Phi_F$) of 82 [85]. The hybrid nanoparticles were then loaded with doxorubicin cargos and the mesoporous particles were finally capped with redox-responsive bis(propyl)disulfide nanogates (Fig. 7A). The nanocaps could be opened by adding a chemical reducer stimulus such as mercaptoethanol, while light actuation in solution required a sacrificial ethylenediaminetetraacetic acid (EDTA) agent to prevent back-electron transfer (Fig. 7B). The release of rhodamine B cargos was also demonstrated upon two-photon excitation at 800 nm as well as in MCF-7 breast cancer cells.

Using the same photo-electron transfer mechanism with a mono-alkoxysilylated analog of the photosensitizer, Zink et al. fabricated β-cyclodextrin-adamantane-disulfide-silica nanocarriers for two-photon-responsive cargo release. The surface functionalization was validated through

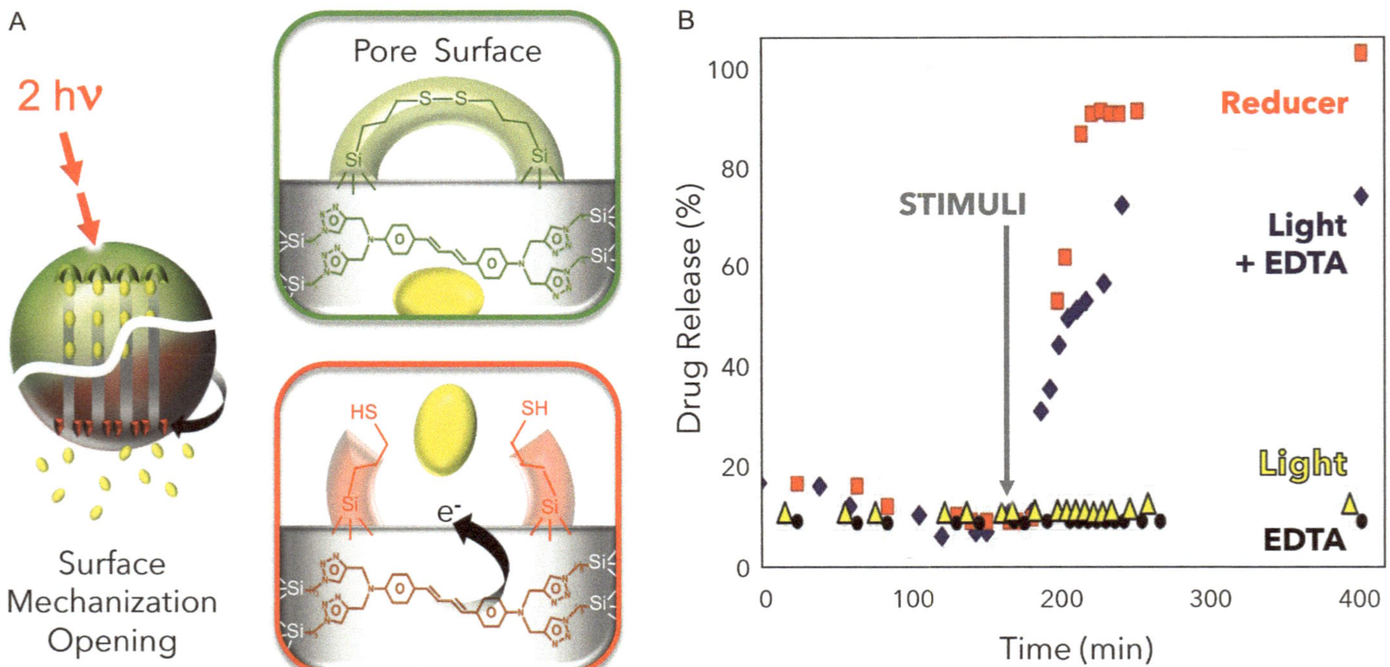

Fig. 7 Representation of the materials design of photosensitizer-doped mesoporous silica nanoparticles surface-engineered for TPA-actuated drug delivery (A). One-photon-actuated doxorubicin release profile upon various stimuli including mercaptoethanol reducers and ethylenediaminetetraacetic acid (EDTA) sacrificial agents to prevent back-electron transfer. Two-photon-actuated release of rhodamine B was also recorded in solution. *Illustration based on the finding of J.G. Croissant, C. Qi, O. Mongin, V. Hugues, M. Blanchard-Desce, L. Raehm, X. Cattoen, M. Wong Chi Man, M. Maynadier, M. Gary-Bobo, M. Garcia, J.I. Zink, J.-O. Durand, Disulfide-gated mesoporous silica nanoparticles designed for two-photon-triggered drug release and imaging, J. Mater. Chem. B, 3 (2015) 6456–6461.*

the releases of caps and cargos upon one- and two-photon light exposures in aqueous media in the presence of EDTA sacrificial agents [69].

3.3 Silica-Fluorophore-Azobenzene Nanocomposites

Our group purposed to configure azobenzene absorbers for two-photon-triggered drug delivery. Azobenzenes can continually photoisomerize from the *cis* to *trans* isomers when exposed to light at a wavelength of the isosbestic point [101–107]. However, this point is in the blue region of the visible spectrum which is non-biomedically suited as pointed out earlier. We thus developed two kinds of two-photon fluorophores which would effectively absorb NIR light under pulsed laser excitations in order to photo-isomerize azobenzenes by a Förster resonance energy transfer (FRET) mechanism [108–111]. The first materials design involved the modification of the pore wall of mesoporous silica nanoparticles with azobenzene pending groups physically trapping drugs, and di-alkoxysilylated fluorophore having a diacetylene-dialkoxy-phenyl core (Fig. 8A) [39]. The presence of the so-called azobenzene nanoimpellers, termed as such for their photoisomerization motion at the nanoscale, leads to a 99% efficient energy transfer upon pulse laser exposure (Fig. 8B). The motion of the nanoimpeller also leads to the expulsion of loaded drugs (Fig. 8A) for the first demonstration of a two-photon-triggered delivery of therapeutic drugs in breast cancer cells (Fig. 8C). Note that using various ratios of azobenzenes over two-photon absorbers, various energy transfers, and in vitro killing efficiencies could be correlated [39]. The $\sigma_2^{700\ \mathrm{nm}}$ in the particles reached up to 300 GM with FRET quantum yield up to 99%.

The second materials design involved the preparation of another two-photon-sensitive fluorophore for FRET toward surface-engineered azobenzene@β-cyclodextrin nanovalves for drug delivery and bioimaging (Fig. 9A) [7]. The fluorophore consisted of di-alkoxysilylated absorbers having a para-cyclophane conjugated core. The particles without the nanovalves surface covering displayed an intense two-photon-actuated fluorescence which was transferred to surface azobenzene valves for effective cancer treatment by light-triggered drug delivery in MCF-7 cells [7]. This design displayed virtually zero passive drug leakage and could be used for both bioimaging and drug delivery since fluorophores located at the center of the particles could not transfer their energy to the surface by a FRET mechanism.

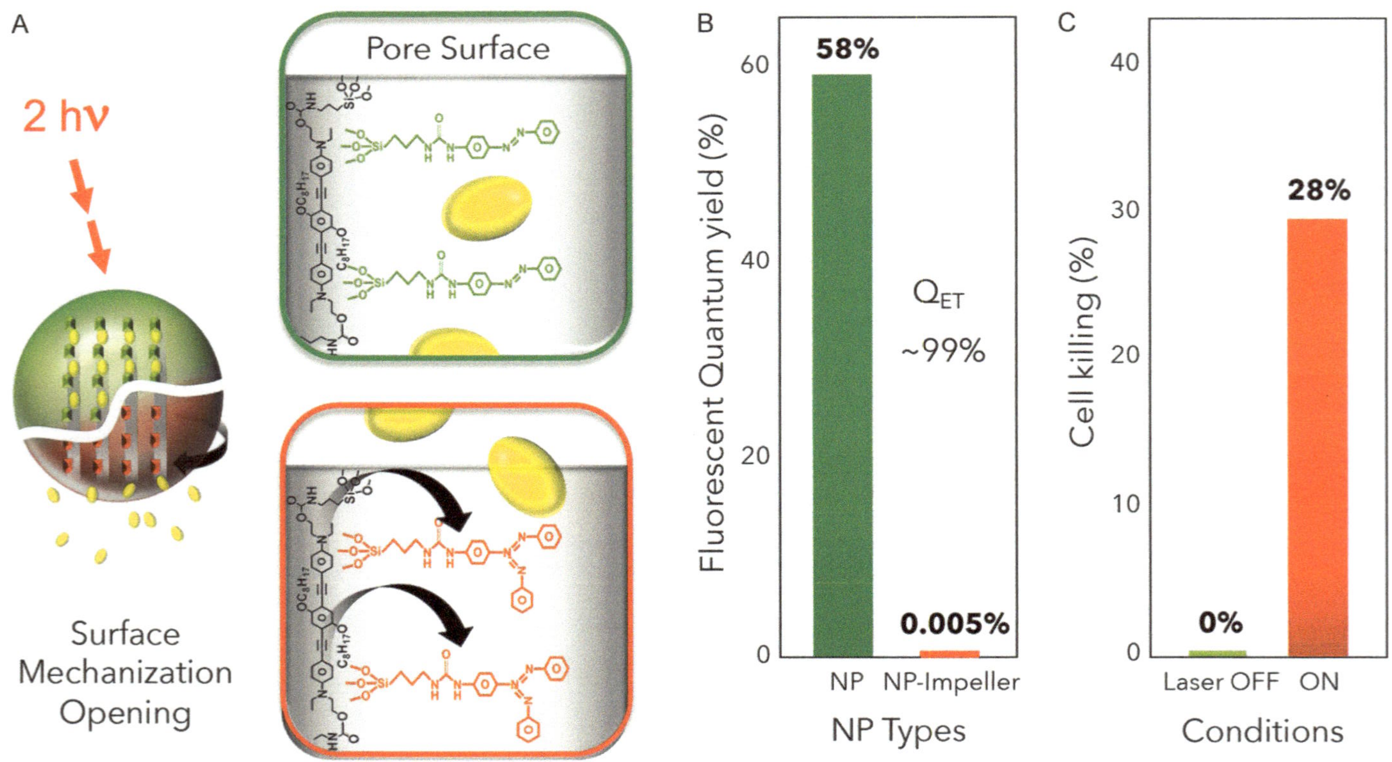

Fig. 8 Representation of the materials design of surface-mechanized mesoporous silica nanoparticles TPA-actuated drug delivery by isomerisation of azobenzene derivatives upon laser exposure (A). The particle pore walls contain two-photon absorbers, covalently linked to silica, and azobenzene nanoimpellers. Fluorescence quantum yield of the particles without and with surface mechanization (NP and NP-Impeller, respectively) (B). MCF-7 cancer cell killing upon two-photon-actuated drug delivery [39]. *Illustration based on the finding of J. Croissant, M. Maynadier, A. Gallud, H. Peindy N'Dongo, J.L. Nyalosaso, G. Derrien, C. Charnay, J.-O. Durand, L. Raehm, F. Serein-Spirau, N. Cheminet, T. Jarrosson, O. Mongin, M. Blanchard-Desce, M. Gary-Bobo, M. Garcia, J. Lu, F. Tamanoi, D. Tarn, T.M. Guardado-Alvarez, J.I. Zink, Two-photon-triggered drug delivery in cancer cells using nanoimpellers, Angew. Chem. Int. Ed. 125 (2013) 14058–14062.*

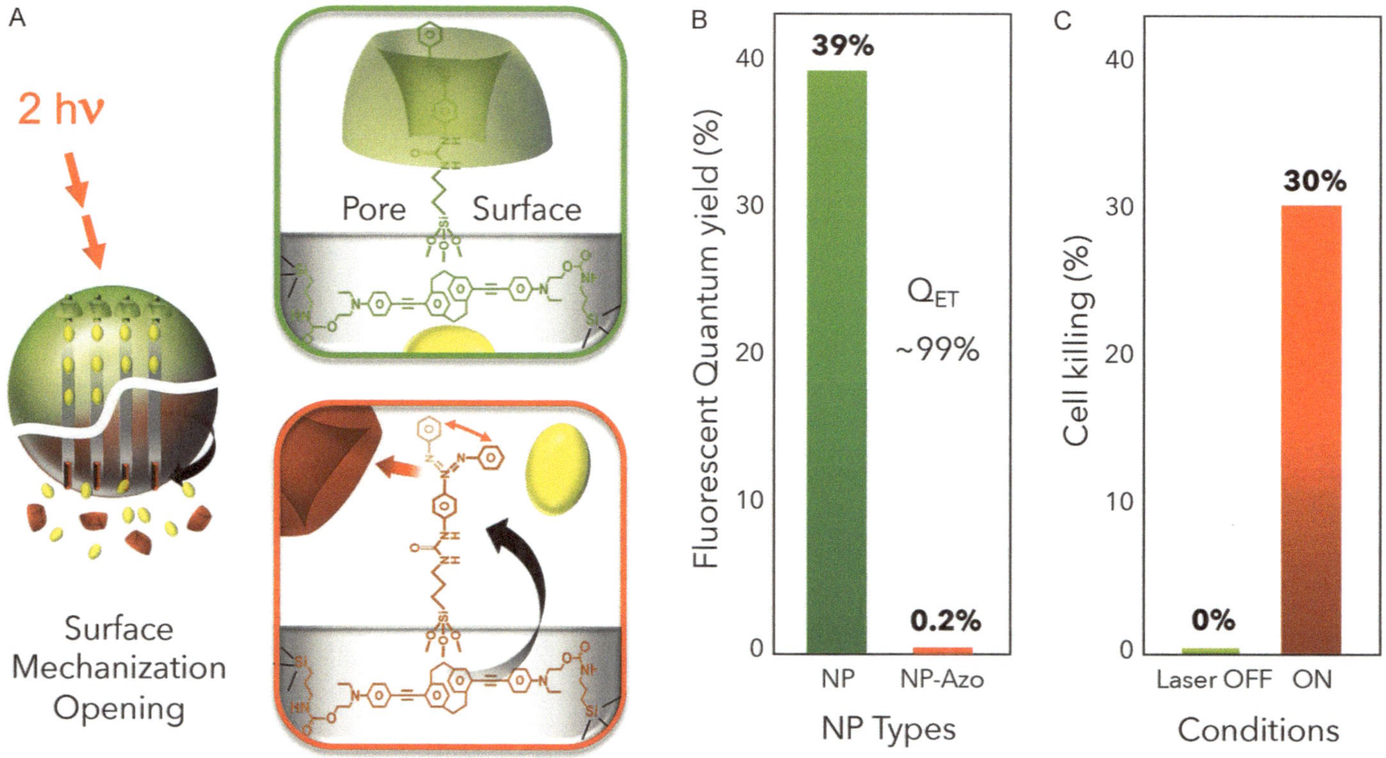

Fig. 9 Representation of the materials design of surface-mechanized mesoporous silica nanoparticles TPA-actuated drug delivery by isomerisation of azobenzene derivatives upon laser exposure (A). The particle pore walls contain two-photon absorbers, covalently linked to silica, and azobenzene@β-cyclodextrin complexes. Fluorescent quantum yield of the particles without and with surface mechanization (NP and NP-Azo, respectively) (B). MCF-7 cancer cell killing upon two-photon-actuated drug delivery [7]. *Illustration based on the finding of J. Croissant, A. Chaix, O. Mongin, M. Wang, S. Clément, L. Raehm, J.-O. Durand, V. Hugues, M. Blanchard-Desce, M. Maynadier, A. Gallud, M. Gary-Bobo, M. Garcia, J. Lu, F. Tamanoi, D.P. Ferris, D. Tarn, J.I. Zink, Two-photon-triggered drug delivery via fluorescent nanovalves, Small 10 (2014) 1752–1755.*

4. TWO-PHOTON-EXCITED PHOTODYNAMIC THERAPY

Photodynamic therapy is a treatment that uses photo-responsive drugs, called photosensitizers, to kill cancer cells by producing electron upon light stimulus which in turn reacts with oxygen or biomolecules to generate cytotoxic reactive oxygen production (ROS) [24,89,112]. As mentioned earlier, most photosensitizers are not biologically stable and cannot penetrate the cell membrane which is why they are often vectorized via nanoparticle formulations [92,113–117]. Photodynamic therapy using mesoporous silica particles was first reported by the group of Mou with the covalent incorporation of protoporphyrin IX for both the time-dependent and the dosage-dependent HeLa cell death upon on-photon laser exposure [118]. In this section, we consider four types of silica nanoformulations: (1) silica-photosensitizer, (2) silica-fluorophore-photosensitizer, (3) silica-gold-photosensitizer, and (4) silica-diamond nanocomposites.

4.1 Silica-Photosensitizer-Nanocomposites

Gary-Bobo et al. reported covalently encapsulated potent photosensitizers into mesoporous silica nanocarriers for the targeted direct photosensitization in vitro and in vivo upon two-photon excitation in the NIR (Fig. 10) [94]. The conjugated photosensitizers had an excellent cross-section of about 1200 GM per chromophore and were excited at 760 nm. The nanoparticle formulation had a hydrodynamic diameter of 118 nm, a σ_2 of up to 8 MGM (per particle), and was surface engineered with mannose targeting moieties. A complete cell death was obtained upon irradiation (Fig. 10B) and the effect was found to be more effective with the targeting agents. In vivo experiment on nude mice treated with nanoparticles depicted up to 70% of tumor weight loss by screening three tumor areas for 3 min (Fig. 10C). Importantly, no metastasis was found which demonstrated the viability of two-photon-triggered photodynamic therapy for real life treatments of cancer pathologies.

Since this groundbreaking work, more studies were published on two-photon-actuated direct photodynamic therapy via silica particles [24,89]. Qian et al. showed that protoporphyrin IX photosensitizer-encapsulated organically modified 25 nm-wide silica nanoparticles could be used for direct two-photon photodynamic therapy and in vivo functional imaging [119]. In a different study, the fabrication of porphyrin- or phthalocyanine-bridged silsesquioxane nanotheranostics for two-photon imaging, photodynamic

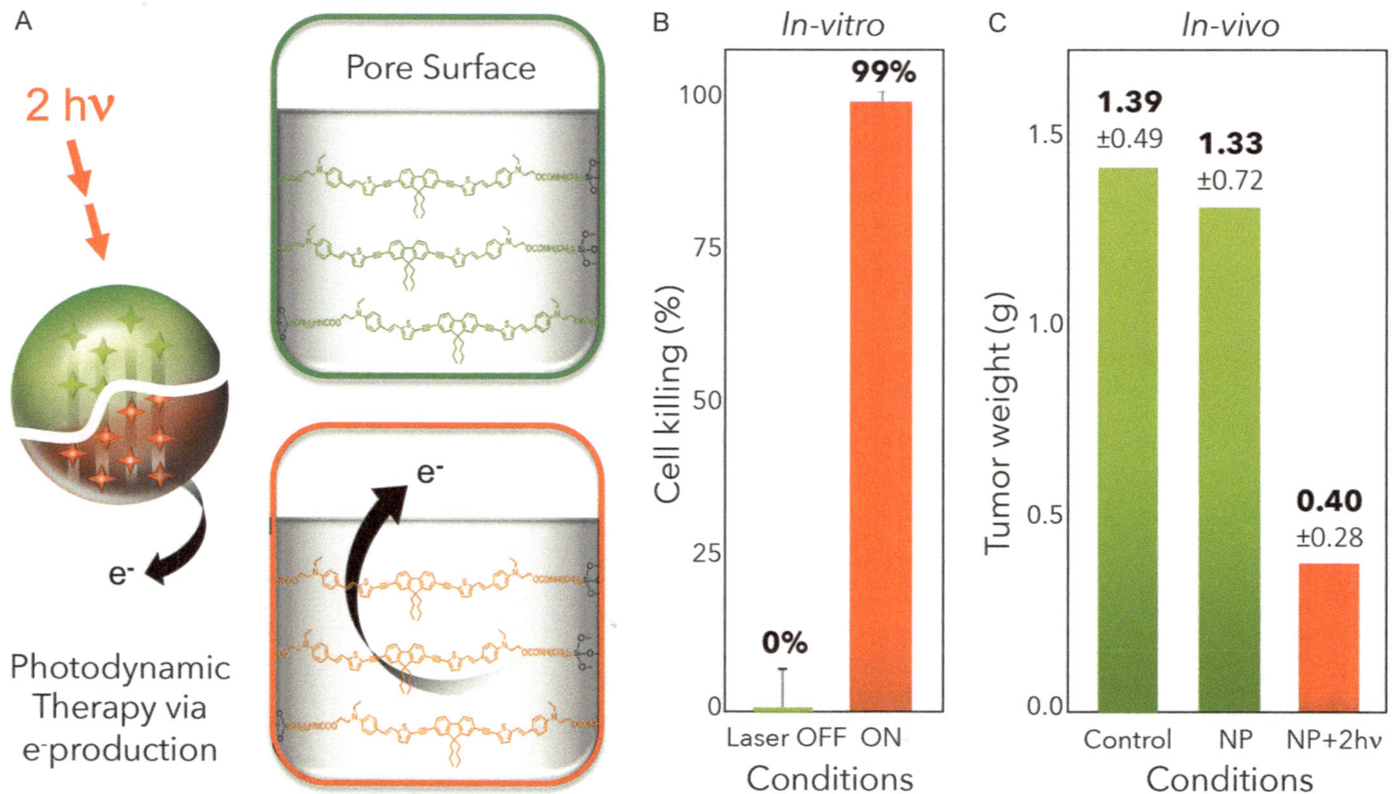

Fig. 10 Representation of mesoporous silica nanoparticles doped with a potent photosensitizer for two-photon-triggered photodynamic therapy (A) in cancer cells (B) and mice (C). *Illustration based on the finding of reference M. Gary-Bobo, Y. Mir, C. Rouxel, D. Brevet, I. Basile, M. Maynadier, O. Vaillant, O. Mongin, M. Blanchard-Desce, A. Morère, M. Garcia, J.-O. Durand, L. Raehm, Mannose-functionalized mesoporous silica nanoparticles for efficient two-photon photodynamic therapy of solid tumors, Angew. Chem. Int. Ed. 123 (2011) 11627–11631.*

therapy, or photoacoustic imaging was demonstrated [120]. The nanoformulation originated from Zn-porphyrin or phthalocyanine tetraalkoxysilanes, depicted a hydrodynamic diameter of about 200 nm, and functioned efficiently against breast cancer cells. In vivo photoacoustic imaging was also carried out on nude mice.

4.2 Silica-Flurophore-Photosensitizer-Nanocomposites

The photosensitization of encapsulated two-photon absorbers can also be triggered indirectly by energy transfer from fluorophores. Prasad and co-workers published the non-porous organosilica nanospheres containing two-photon-absorbing anthracene-based fluorophores as well as pyropheophorbide-based photosensitizers [121]. The chromophores used for the energy transfer possessed a $\sigma_2^{775\,\mathrm{nm}}$ of 217 GM and the hybrid-silica particles effectively generated ROS and photodynamic therapy. The limitation of the remarkable study was the non-covalent doping of the photoabsorbers.

FRET transfer was also reported on covalently linked *Silica-Fluorophore-Photosensitizers Nanocomposites by various groups*. One approach consisted in using fluorescein dyes ($\sigma_2^{782\,\mathrm{nm}} = 38\,\mathrm{GM}$) incorporated into the pore walls of mesoporous organosilica nanoparticles in order to favor the FRET toward Palladium-porphyrins at 900 nm [122]. The report demonstrated that by tailoring the ratio between the two-photon donor antennas and photosensitizer acceptors, the energy transfer quantum yields could be varied (up to 93%). Photodynamic therapy applications in vitro and in vivo were obtained against breast cancers. Using the same principle of FRET between fluorescein isothiocyanate linked to hyperbranched polyglycerol and rose bengal photosensitizer trapped in the pores of mesoporous silica nanoparticles, Liu et al. achieved apoptosis of HeLa cells with two-photon excitation at 920 nm [123]. Our group and the group of Blanchard-Desce recently synthesized a large and potent two-photon fluorophore ($\sigma_2^{740\,\mathrm{nm}} = 1100\,\mathrm{GM}$) for FRET-mediated photodynamic therapy with more sensitivity. The sub-100 nm porous silica hybrid particles contained both the fluorophore acceptor and photosensitizer antennas inside the framework, attached by covalent linkages. In addition, the large porosity of the nanocarriers allowed the loading and autonomous pH-triggered release of anticancer drugs. The therapeutic photo-induced effectiveness of the formulation was demonstrated on cancer cells at low power (3%, input power 3 W, 760 nm) thanks to unusually high $\sigma_2^{740\,\mathrm{nm}}$ per chromophore (~2300 GM) [124].

4.3 Silica-Gold-Photosensitizer Nanocomposites

Gold plasmonic core [125,126] can be used both to produce electron for directly photodynamic therapy, and to transfer energy to indirectly excite photosensitizers [1]. The direct production of ROS by gold particles has been recently established for citrate-stabilized gold nanospheres [127], gold nanoclusters [128], gold nanorods [129], gold nanocages [130], and aggregated of spherical and rod-shaped gold nanomaterials [131]. The direct and indirect mechanisms by which gold nanostructures can be instrumental for photo-induced therapies come from their variously high two-photon cross-sections ranging from a few thousand GM in gold nanoclusters to σ_2^{MAX} of approximately 100,000 GM for nanorods [132].

Chen and co-workers fabricated gold nanorod core@mesoporous silica-porphyrin shell nanoparticles for the plasmonic FRET-mediated electron production and photodynamic therapy (Fig. 11) [133]. An average power of about 100 mW was applied on a 7 mm circular area (leading to ~250 mW cm^{-2}) to excite the plasmonic cores. The singlet oxygen production by porphyrin guests adsorbed into the mesoporous shell was triggered in vivo at 800, in spite of despite a σ_2 of only ~2 GM ($^1O_2$ quantum yield of 0.47) due to the effective FRET from the gold nanorods. A five-fold increase of the tumor inhibition was observed with the silica-gold-porphyrin nanocomposites (Fig. 11B). Croissant et al. demonstrated the exaltation of the photodynamic therapy potential of diphenylbutadiene photosensitizers using aggregated gold nanosphere@bridged silsesquioxane core-shell nanoparticles [91]. Both the photophysical properties and the biological therapeutic efficacies were enhanced by the presence of gold cores (as well as surface decorated gold nanospheres). The cross-sections of the molecular photosensitizer, photosensitizer-based particles, and gold@photosensitizer-based particles were respectively of 100, 140, and 260 GM, due to an exaltation of the TPA via plasmonic FRET. The advantages of using gold antennas include the possibility to tune the FRET by changing the size and morphologies of the gold nanostructures, or by varying the thickness of the silica hybrid shell since the gold-photosensitizer distance impacts the therapeutic effect.

4.4 Silica-Diamond Nanocomposites

A direct electron production to generate ROS was also reported for the first time by Durand et al. using nanodiamond [134] core@mesoporous silsesquioxane nanoparticles [96]. The hydroxyl-terminated nanodiamond cores exhibited a primary particle size of 4.0 nm on average and were

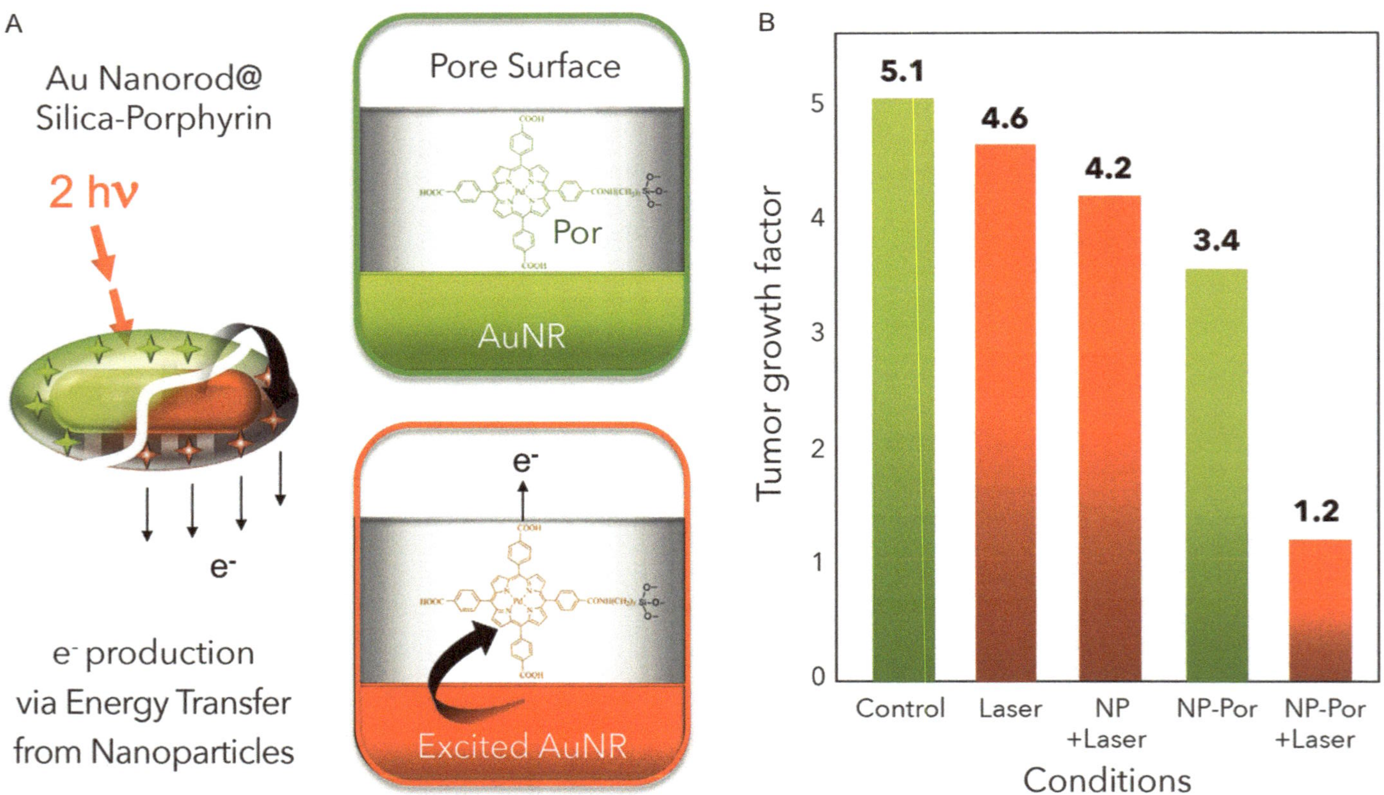

Fig. 11 Representation of the gold nanorod core@mesoporous silica-porphyrin shell nanoparticles for plasmonic FRET-mediated electron production and photodynamic therapy (A). Tumor growth factor difference before and after laser irradiation in various conditions (B). *Illustration based on the finding of N.-T. Chen, K.-C. Tang, M.-F. Chung, S.-H. Cheng, C.-M. Huang, C.-H. Chu, P.-T. Chou, J.S. Souris, C.-T. Chen, C.-Y. Mou, L.-W. Lo, Enhanced plasmonic resonance energy transfer in mesoporous silica-encased gold nanorod for two-photon-activated photodynamic therapy, Theranostics 4 (2014) 798–807.*

aggregated into few tens of nanometers before being encapsulated in the hybrid siliceous nanoparticles. The detection of intracellular ROS in MCF-7 cancer cells was carried out using the DCFDA kit (i.e., dichlorodihydrofluorescein diacetate). Cells were incubated with a concentration of nanoparticles of $80\,\mu g\,mL^{-1}$ and the two-photon absorption of the nanodiamonds (800 nm, $3 \times 1.57\,s$) was monitored by green luminescence to produce ROS. Nearly 40% of selective cancer cell death was obtained using that methodology. The nanoplatform was also used for synergistic drug delivery and photodynamic therapy (see Section 5 hereafter).

5. TWO-PHOTON-EXCITED MULTI-THERAPY

5.1 Combining Drug Delivery and Photodynamic Therapy

Durand et al. designed silica-based nanoparticles covalently doped with a *p*-sulfonatophenyl-porphyrin photosensitizers, surface-functionalized with galactose, and loaded camptothecin anticancer drugs for multi-therapy of cancer [135]. The multifunctional nanocarriers were then applied for targeted drug delivery and photodynamic therapy against living HCT-116 colorectal, MDA-MB-231 human breast adenocarcinoma, and Capan-1 pancreatic cancer cells. The authors established that such cells have a high affinity for glycosylated targeting and obtained 80 to 100% cell killing. This was first demonstration of the synergic effect of two active principles (i.e., porphyrin photosensitizers and camptothecin drugs) in the same lectin-targeted silica nanocarriers [135].

Nanoparticles with bridged silsesquioxane-based frameworks [136–140] were applied for two-photon-actuated multi-therapy involving a combination of drug delivery and photodynamic therapy [141]. Mesoporous bridged silsesquioxane nanoparticles [142] have unique advantages arising from their medium to high organic content (20–80 wt% [22,142]) including: (i) tunable hydrophobicity/hydrophilicity [143–145], (ii) exceptional cargo loading capacities (up to 1200 mg of cargo per gram of nanoparticle [136,137]), (iii) virtually zero cargo leakage without pore capping and a pH-triggered release reported for certain drugs [21,146], (iv) tunable degradability rates and mechanisms [136,146,147], and (v) very low hemolytic effects [148,149] (lower than the best silicas [10]) and an excellent biocompatibility [10]. Using ethenylene-bridged and phenylene-bridged mesoporous silsesquioxane nanoparticles covalently doped with high contents of diphenylbutadiene-based photosensitizers (~25 wt%), Croissant et al. showed the combination of photodynamic therapy and drug delivery in cancer cells. A library of spherical nanomaterials was described with monodisperse diameters of 100–200 nm and high surface areas of about 1000 $m^2\ g^{-1}$. The unloaded nanoparticles demonstrated ca. 30–40% of cell killing upon two-photon excitation at 760 nm by three scans of 1.57 s duration. Both hydrophobic camptothecin and hydrophilic doxorubicin drugs could be stably loaded and generated 30–45% of cell death by autonomously delivery the drugs to cells. In control experiments utilizing the nanocarriers without photosensitizer doping, turning the pulsed laser ON presumably increased the delivery capability of the nanocarriers and thereby increased the observed cell death. Combining the loading of anticancer drug in photosensitizer-doped nanoparticles led to synergistic drug delivery and photodynamic therapy with up to 76% of cancer cell death. These multifunctional particles could also be tracked via their two-photon fluorescence [141].

The same group later reported the synergistic delivery of doxorubicin anticancer drug in combination with photodynamic therapy arising from nanodiamond core@ethenylene-bridged silsesquioxane shell nanocarriers [96]. The particles have surface areas of ca. ~900 to over 1400 $m^2\ g^{-1}$ and could host up to 70 wt% of doxorubicin. The drug could be autonomously and effectively released and delivered in acidic conditions (~pH 5.0). The nanosystems were applied for combination therapy and afforded up to 87% cell killing upon irradiation at 800 nm.

5.2 Combining Photodynamic and Gene Silencing Therapy

We recently published a novel nanoparticle fully composed of giant organic bridges based on octasilylated metalated porphyrins resulting in an unprecedented ~80 wt% of organic content in a mesoporous nanomaterial (Fig. 12A) [22]. Despite this high content, which generally leads to the absence of mesopores [138], a surface area of 230 $m^2\ g^{-1}$ was calculated which was consistent with the wide pore distribution from 4 to 80 nm of the nanoparticles which displayed interconnected cavities and hollow-like nanostructure as shown by the transmission electron microscopy (TEM) image in Fig. 12B. The octasilylated porphyrin groups in the pore wall of the nanoparticles were analyzed by a battery of techniques such as high-resolution TEM, UV-vis and fluorescence spectroscopies, and density functional theory (DFT) studies which showed unique J-aggregates of porphyrins which are known to harness TPA [63,150]. For instance, UV-vis spectroscopy displays an intense Soret band at 439 nm and two red-shifted

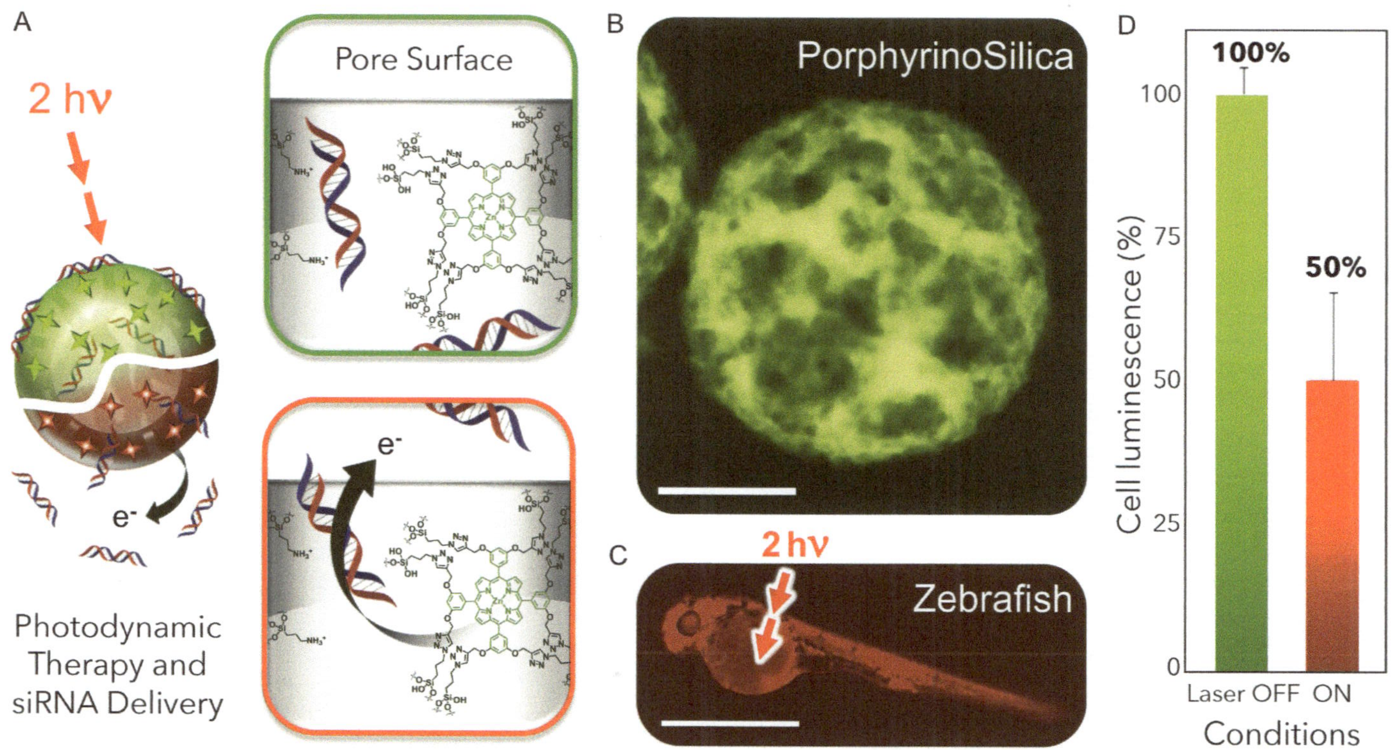

Fig. 12 Representation of the materials design of porphyrinosilica nanoparticles for siRNA delivery to zebrafish and TPA-actuated photodynamic therapy upon laser exposure (A). Artificially colored TEM image of the porphyrinosilica nanoparticles showing their interconnected cavity-like hollow structure (B). Artificially colored photograph of a zebrafish (the arrows indicate the location of the pulsed laser irradiation) (C). Cell fluorescence results before and after TPA indicate an effective gene therapy only upon laser exposure, which strongly suggests a photochemical internalization (PCI) of siRNAs.

Q bands at 561 and 604 nm which are indicative of J-aggregates of porphyrins. This property was demonstrated on MDA-MB-231 breast cancer cells which were incubated with the nanoparticles and generated 73% of cytotoxicity upon two-photon excitation at 800 nm. In addition, the silsesquioxane particles were engineered with aminopropyl groups to electrostatically bind siRNA moieties that could be incorporated for gene silencing experiments in zebrafish (Fig. 12A and C). TPE photochemical internalization (PCI) of siRNAs was demonstrated with targeting of the luciferase genes in MCF-7 cancer cells expressing luciferase which afforded to 50% of gene silencing therapy (Fig. 12D).

6. OUTLOOK

In this chapter, we show that using two-photon near-infrared light for the spatio-temporal control of theranostics is a very promising route particularly for the deep treatment of tumors. The features of two-photon light would allow imaging and therapy of small-sized tumors detected at an early stage of the disease by screening campaigns, where classical invasive treatments (radiotherapy, chemotherapy, surgery) are not adapted. The recent progress of chemistry and physics allows designing very efficient two-photon absorbers and photosensitizers. The diversity of nanoparticle formulations designed for two-photon excitation leads to versatile two-photon-triggered systems for the delivery of reactive oxygen species, drugs and genes in order to efficiently kill cancer cells in vitro and in vivo. Many cancers could be targeted with two-photon excitation (retinoblastoma, prostate cancer, breast cancer, colon cancer...), indeed two-photon endoscopes exist [151] and new formulations combined with these endoscopes are needed for the precise medicine of the future. The diversity of the formulations and treatments available with two-photon could be selected and adapted depending on the cancer type in order to obtain the best effects in terms of theranostics. Therefore, two-photon excitation would be particularly adapted for the spatio-temporal personalized medicine of the future.

ACKNOWLEDGMENTS

We acknowledge support by the Partner University Fund 20091853, ANRs Mechanano, NanoptPDT.

REFERENCES

[1] J.G. Croissant, J.I. Zink, L. Raehm, J.O. Durand, Two-photon-excited silica and organosilica nanoparticles for spatiotemporal cancer treatment, Adv. Healthc. Mater. 7 (2018) 1701248.

[2] S. Mura, J. Nicolas, P. Couvreur, Stimuli-responsive nanocarriers for drug delivery, Nat. Mater. 12 (2013) 991–1003.

[3] L. Brannon-Peppas, J.O. Blanchette, Nanoparticle and targeted systems for cancer therapy, Adv. Drug Deliv. Rev. 64 (2012) 206–212.

[4] K.H. Bae, H.J. Chung, T.G. Park, Nanomaterials for cancer therapy and imaging, Mol. Cells 31 (2011) 295–302.

[5] D. Bechet, P. Couleaud, C. Frochot, M.-L. Viriot, F. Guillemin, M. Barberi-Heyob, Nanoparticles as vehicles for delivery of photodynamic therapy agents, Trends Biotechnol. 26 (2008) 612–621.

[6] P. Juzenas, W. Chen, Y.P. Sun, M.A.N. Coelho, R. Generalov, N. Generalova, I.L. Christensen, Quantum dots and nanoparticles for photodynamic and radiation therapies of cancer, Adv. Drug Deliv. Rev. 60 (2008) 1600–1614.

[7] J. Croissant, A. Chaix, O. Mongin, M. Wang, S. Clément, L. Raehm, J.-O. Durand, V. Hugues, M. Blanchard-Desce, M. Maynadier, A. Gallud, M. Gary-Bobo, M. Garcia, J. Lu, F. Tamanoi, D.P. Ferris, D. Tarn, J.I. Zink, Two-photon-triggered drug delivery via fluorescent nanovalves, Small 10 (2014) 1752–1755.

[8] C. Conte, F. Ungaro, A. Mazzaglia, F. Quaglia, Photodynamic therapy for cancer: principles, clinical applications, and nanotechnological approaches, in: Nano-oncologicals, Springer, 2014, pp. 123–160.

[9] P. Agostinis, K. Berg, K.A. Cengel, T.H. Foster, A.W. Girotti, S.O. Gollnick, S.M. Hahn, M.R. Hamblin, A. Juzeniene, D. Kessel, Photodynamic therapy of cancer: an update, CA Cancer J. Clin. 61 (2011) 250–281.

[10] J.G. Croissant, Y. Fatieiev, A. Almalik, N.M. Khashab, Mesoporous silica and organosilica nanoparticles: physical chemistry, biosafety, delivery strategies, and biomedical applications, Adv. Healthc. Mater. 7 (2018) 1700831.

[11] S. Li, B.A. Moosa, J.G. Croissant, N.M. Khashab, Electrostatic assembly/disassembly of nanoscaled colloidosomes for light-triggered cargo release, Angew. Chem. Int. Ed. 127 (2015) 6908–6912.

[12] Y. Fatieiev, J. Croissant, S. Alsaiari, B.A. Moosa, D.H. Anjum, N.M. Khashab, Photoresponsive bridged silsesquioxane nanoparticles with tunable morphology for light-triggered plasmid DNA delivery, ACS Appl. Mater. Interfaces 7 (2015) 24993–24997.

[13] J. Croissant, J.I. Zink, Nanovalve-controlled cargo release activated by plasmonic heating, J. Am. Chem. Soc. 134 (2012) 7628–7631.

[14] Y. Wang, Q. Zhao, N. Han, L. Bai, J. Li, J. Liu, E. Che, L. Hu, Q. Zhang, T. Jiang, S. Wang, Mesoporous silica nanoparticles in drug delivery and biomedical applications, Nanomedicine 11 (2015) 313–327.

[15] W. Ji, N. Li, D. Chen, X. Qi, W. Sha, Y. Jiao, Q. Xu, J. Lu, Coumarin-containing photo-responsive nanocomposites for NIR light-triggered controlled drug release via a two-photon process, J. Mater. Chem. B 1 (2013) 5942–5949.

[16] K.K. Coti, M.E. Belowich, M. Liong, M.W. Ambrogio, Y.A. Lau, H.A. Khatib, J.I. Zink, N.M. Khashab, J.F. Stoddart, Mechanised nanoparticles for drug delivery, Nanoscale 1 (2009) 16–39.

[17] Y. Yang, F. Liu, X. Liu, B. Xing, NIR light controlled photorelease of siRNA and its targeted intracellular delivery based on upconversion nanoparticles, Nanoscale 5 (2013) 231–238.

[18] X. Du, X. Li, L. Xiong, X. Zhang, F. Kleitz, S.Z. Qiao, Mesoporous silica nanoparticles with organo-bridged silsesquioxane framework as innovative platforms for bioimaging and therapeutic agent delivery, Biomaterials 91 (2016) 90–127.
[19] L. Zhang, E. Wang, Metal nanoclusters: new fluorescent probes for sensors and bioimaging, Nano Today 9 (2014) 132–157.
[20] I. Roy, P. Kumar, R. Kumar, T.Y. Ohulchanskyy, K.-T. Yong, P.N. Prasad, Ormosil nanoparticles as a sustained-release drug delivery vehicle, RSC Adv. 4 (2014) 53498–53504.
[21] D. Aggad, C.M. Jimenez, S. Dib, J.G. Croissant, L. Lichon, D. Laurencin, S. Richeter, M. Maynadier, S.K. Alsaiari, M. Boufatit, Gemcitabine delivery and photodynamic therapy in cancer cells via porphyrin-ethylene-based periodic mesoporous organosilica nanoparticles, ChemNanoMat 4 (2018) 46–51.
[22] C.M. Jimenez, D. Aggad, J.G. Croissant, K. Tresfield, D. Laurencin, D. Berthomieu, N. Cubedo, M. Rossel, S. Alsaiari, D.H. Anjum, R. Sougrat, M.A. Roldan-Gutierrez, S. Richeter, E. Oliviero, L. Raehm, C. Charnay, X. Cattoën, S. Clément, M.W.C. Man, M. Maynadier, V. Chaleix, V. Sol, M. Garcia, M. Gary-Bobo, N.M. Khashab, N. Bettache, J.-O. Durand, Porous porphyrin-based organosilica nanoparticles for NIR two-photon photodynamic therapy and gene delivery in zebrafish, Adv. Funct. Mater. 28 (2018) 1800235.
[23] B. Sun, L. Wang, Q. Li, P. He, H. Liu, H. Wang, Y. Yang, J. Li, Bis (pyrene)-doped cationic dipeptide nanoparticles for two-photon-activated photodynamic therapy, Biomacromolecules 18 (2017) 3506–3513.
[24] Y. Shen, A.J. Shuhendler, D. Ye, J.-J. Xu, H.-Y. Chen, Two-photon excitation nanoparticles for photodynamic therapy, Chem. Soc. Rev. 45 (2016) 6725–6741.
[25] M. Firczuk, M. Winiarska, A. Szokalska, M. Jodlowska, M. Swiech, K. Bojarczuk, P. Salwa, D. Nowis, Approaches to improve photodynamic therapy of cancer, Front. Biosci. 16 (2011) 208–224.
[26] O. Hocine, M. Gary-Bobo, D. Brevet, M. Maynadier, S. Fontanel, L. Raehm, S. Richeter, B. Loock, P. Couleaud, C. Frochot, C. Charnay, G. Derrien, M. Smaihi, A. Sahmoune, A. Morere, P. Maillard, M. Garcia, J.-O. Durand, Silicalites and mesoporous silica nanoparticles for photodynamic therapy, Int. J. Pharm. 402 (2010) 221–230.
[27] B.C. Wilson, Photodynamic therapy/diagnostics: principles, practice, and advances, Handb. Photonics Biomed. Sci. (2010) 649–686.
[28] P. Couleaud, V. Morosini, C. Frochot, S. Richeter, L. Raehm, J.-O. Durand, Silica-based nanoparticles for photodynamic therapy applications, Nanoscale 2 (2010) 1083–1095.
[29] J. Wu, Y. Liu, Y. Tang, S. Wang, C. Wang, Y. Li, X. Su, J. Tian, Y. Tian, J. Pan, Y. Su, H. Zhu, Z. Teng, G. Lu, Synergistic chemo–photothermal therapy of breast cancer by mesenchymal stem cell-encapsulated yolk–shell GNR@HPMO-PTX nanospheres, ACS Appl. Mater. Interfaces 8 (2016) 17927–17935.
[30] X. Tu, L. Wang, Y. Cao, Y. Ma, H. Shen, M. Zhang, Z. Zhang, Efficient cancer ablation by combined photothermal and enhanced chemo-therapy based on carbon nanoparticles/doxorubicin@ SiO_2 nanocomposites, Carbon 97 (2016) 35–44.
[31] S. Shen, H. Tang, X. Zhang, J. Ren, Z. Pang, D. Wang, H. Gao, Y. Qian, X. Jiang, W. Yang, Targeting mesoporous silica-encapsulated gold nanorods for chemo-photothermal therapy with near-infrared radiation, Biomaterials 34 (2013) 3150–3158.
[32] X. Xue, F. Wang, X. Liu, Emerging functional nanomaterials for therapeutics, J. Mater. Chem. 21 (2011) 13107–13127.
[33] X. Huang, W. Qian, I.H. El-Sayed, M.A. El-Sayed, The potential use of the enhanced nonlinear properties of gold nanospheres in photothermal cancer therapy, Lasers Surg. Med. 39 (2007) 747–753.

[34] Y. Wu, C. Chen, S. Liu, Enzyme-functionalized silica nanoparticles as sensitive labels in biosensing, Anal. Chem. 81 (2009) 1600–1607.
[35] X. He, H. Nie, K. Wang, W. Tan, X. Wu, P. Zhang, In vivo study of biodistribution and urinary excretion of surface-modified silica nanoparticles, Anal. Chem. 80 (2008) 9597–9603.
[36] I.I. Slowing, B.G. Trewyn, S. Giri, V.S.Y. Lin, Mesoporous silica nanoparticles for drug delivery and biosensing applications, Adv. Funct. Mater. 17 (2007) 1225–1236.
[37] M. Hartmann, Ordered mesoporous materials for bioadsorption and biocatalysis, Chem. Mater. 17 (2005) 4577–4593.
[38] D. Moelans, P. Cool, J. Baeyens, E.F. Vansant, Using mesoporous silica materials to immobilise biocatalysis-enzymes, Cat. Com. 6 (2005) 307–311.
[39] J. Croissant, M. Maynadier, A. Gallud, H. Peindy N'Dongo, J.L. Nyalosaso, G. Derrien, C. Charnay, J.-O. Durand, L. Raehm, F. Serein-Spirau, N. Cheminet, T. Jarrosson, O. Mongin, M. Blanchard-Desce, M. Gary-Bobo, M. Garcia, J. Lu, F. Tamanoi, D. Tarn, T.M. Guardado-Alvarez, J.I. Zink, Two-photon-triggered drug delivery in cancer cells using nanoimpellers, Angew. Chem. Int. Ed. 125 (2013) 14058–14062.
[40] K.K. Jain, K.K. Jain, The Handbook of Nanomedicine, Springer, 2017.
[41] L. Xu, F. Mou, H. Gong, M. Luo, J. Guan, Light-driven micro/nanomotors: from fundamentals to applications, Chem. Soc. Rev. 46 (2017) 6905–6926.
[42] S. Shen, S. Wang, R. Zheng, X. Zhu, X. Jiang, D. Fu, W. Yang, Magnetic nanoparticle clusters for photothermal therapy with near-infrared irradiation, Biomaterials 39 (2015) 67–74.
[43] X. He, K. Wang, Z. Cheng, In vivo near-infrared fluorescence imaging of cancer with nanoparticle-based probes, Wiley Interdiscip. Rev. Nanomed. Nanobiotechnol. 2 (2010) 349–366.
[44] C.-L. Tsai, J.-C. Chen, W.-J. Wang, Near-infrared absorption property of biological soft tissue constituents, J. Med. Biol. Eng. 21 (2001) 7–14.
[45] H.S. Qian, H.C. Guo, P.C.L. Ho, R. Mahendran, Y. Zhang, Mesoporous-silica-coated up-conversion fluorescent nanoparticles for photodynamic therapy, Small 5 (2009) 2285–2290.
[46] B.I. Kruft, A. Greer, Photosensitization reactions in vitro and in vivo, Photochem. Photobiol. 87 (2011) 1204–1213.
[47] X. Ding, C.H. Liow, M. Zhang, R. Huang, C. Li, H. Shen, M. Liu, Y. Zou, N. Gao, Z. Zhang, Surface plasmon resonance enhanced light absorption and photothermal therapy in the second near-infrared window, J. Am. Chem. Soc. 136 (2014) 15684–15693.
[48] J.R. Starkey, A.K. Rebane, M.A. Drobizhev, F.Q. Meng, A.J. Gong, A. Elliott, K. McInnerney, C.W. Spangler, New two-photon activated photodynamic therapy sensitizers induce xenograft tumor regressions after near-ir laser treatment through the body of the host mouse, Clin. Cancer Res. 14 (2008) 6564–6573.
[49] A.M. Smith, M.C. Mancini, S. Nie, Bioimaging: second window for in vivo imaging, Nat. Nanotechnol. 4 (2009) 710–711.
[50] M. Oheim, D.J. Michael, M. Geisbauer, D. Madsen, R.H. Chow, Principles of two-photon excitation fluorescence microscopy and other nonlinear imaging approaches, Adv. Drug Deliv. Rev. 58 (2006) 788–808.
[51] W. Kaiser, C. Garrett, Two-photon excitation in CaF_2: Eu^{2+}, Phys. Rev. Lett. 7 (1961) 229.
[52] I. Abella, Optical double-photon absorption in cesium vapor, Phys. Rev. Lett. 9 (1962) 453.
[53] E.R. Andresen, G. Bouwmans, S. Monneret, H. Rigneault, Two-photon lensless endoscope, Opt. Express 21 (2013) 20713–20721.

[54] L. Porrès, O. Mongin, C. Katan, M. Charlot, T. Pons, J. Mertz, M. Blanchard-Desce, Enhanced two-photon absorption with novel octupolar propeller-shaped fluorophores derived from triphenylamine, Org. Lett. 6 (2003) 47–50.
[55] M. Blanchard-Desce, Molecular engineering of NLO-phores for new NLO microscopies, C. R. Phys. 3 (2002) 439–448.
[56] M.A. Albota, C. Xu, W.W. Webb, Two-photon fluorescence excitation cross sections of biomolecular probes from 690 to 960 nm, Appl. Optics 37 (1998) 7352–7356.
[57] M. Bijeesh, P. Shakhi, S. Arunkarthick, G.K. Varier, P. Nandakumar, Confocal imaging of single $BaTiO_3$ nanoparticles by two-photon photothermal microscopy, Sci. Rep. 7 (2017) 1643.
[58] W.-S. Kuo, Y.-T. Shao, K.-S. Huang, T.-M. Chou, C.-H. Yang, Antimicrobial amino-functionalized nitrogen-doped graphene quantum dots for eliminating multidrug-resistant species in dual-modality photodynamic therapy and bioimaging under two-photon excitation, ACS Appl. Mater. Interfaces 10 (2018) 14438–14446.
[59] S.M. Ardekani, A. Dehghani, M. Hassan, M. Kianinia, I. Aharonovich, V.G. Gomes, Two-photon excitation triggers combined chemo-photothermal therapy via doped carbon nanohybrid dots for effective breast cancer treatment, Chem. Eng. J. 330 (2017) 651–662.
[60] W. Yang, L. Carrillo-Reid, Y. Bando, D.S. Peterka, R. Yuste, Simultaneous two-photon imaging and two-photon optogenetics of cortical circuits in three dimensions, Elife 7 (2018) e32671.
[61] W. Yang, L.-M. Fu, X. Wen, Y. Liu, Y. Tian, Y.-C. Liu, R.-C. Han, Z.-Y. Gao, T.-E. Wang, Y.-L. Sha, Nanoprobes for two-photon excitation time-resolved imaging of living animals: in situ analysis of tumor-targeting dynamics of nanocarriers, Biomaterials 100 (2016) 152–161.
[62] B.C. Marin, S.-W. Hsu, L. Chen, A. Lo, D.W. Zwissler, Z. Liu, A.R. Tao, Plasmon-enhanced two-photon absorption in photoluminescent semiconductor nanocrystals, ACS Photonics 3 (2016) 526–531.
[63] A. Mikhaylov, D.V. Kondratuk, A. Cnossen, H.L. Anderson, M. Drobizhev, A. Rebane, Cooperative enhancement of two-photon absorption in self-assembled zinc-porphyrin nanostructures, J. Phys. Chem. C 120 (2016) 11663–11670.
[64] H.-M. Meng, Z. Jin, Y. Lv, C. Yang, X.-B. Zhang, W. Tan, R.-Q. Yu, Activatable two-photon fluorescence nanoprobe for bioimaging of glutathione in living cells and tissues, Anal. Chem. 86 (2014) 12321–12326.
[65] L. Bertazza, L. Celotti, G. Fabbrini, M.A. Loi, M. Maggini, F. Mancin, S. Marcuz, E. Menna, M. Muccini, U. Tonellato, Cell penetrating silica nanoparticles doped with two-photon absorbing fluorophores, Tetrahedron 62 (2006) 10434–10440.
[66] Y. Yuan, Z. Wang, P. Cai, J. Liu, L.-D. Liao, M. Hong, X. Chen, N. Thakor, B. Liu, Conjugated polymer and drug co-encapsulated nanoparticles for chemo-and photo-thermal combination therapy with two-photon regulated fast drug release, Nanoscale 7 (2015) 3067–3076.
[67] J.G. Croissant, C. Qi, O. Mongin, V. Hugues, M. Blanchard-Desce, L. Raehm, X. Cattoen, M. Wong Chi Man, M. Maynadier, M. Gary-Bobo, M. Garcia, J.I. Zink, J.-O. Durand, Disulfide-gated mesoporous silica nanoparticles designed for two-photon-triggered drug release and imaging, J. Mater. Chem. B 3 (2015) 6456–6461.
[68] T.M. Guardado-Alvarez, L. Sudha Devi, M.M. Russell, B.J. Schwartz, J.I. Zink, Activation of snap-top capped mesoporous silica nanocontainers using two near-infrared photons, J. Am. Chem. Soc. 135 (2013) 14000–14003.
[69] T.M. Guardado-Alvarez, L.S. Devi, J.-M. Vabre, T.A. Pecorelli, B.J. Schwartz, J.-O. Durand, O. Mongin, M. Blanchard-Desce, J.I. Zink, Photo-redox activated

drug delivery systems operating under two photon excitation in the near-IR, Nanoscale 6 (2014) 4652–4658.
[70] S.S. Banerjee, D.-H. Chen, A multifunctional magnetic nanocarrier bearing fluorescent dye for targeted drug delivery by enhanced two-photon triggered release, Nanotechnology 20 (2009) 185103.
[71] D. Kehrloesser, P.J. Behrendt, N. Hampp, Two-photon absorption triggered drug delivery from a polymer for intraocular lenses in presence of an UV-absorber, J. Photoch. Photobio. A 248 (2012) 8–14.
[72] Q.N. Lin, Q. Huang, C.Y. Li, C.Y. Bao, Z.Z. Liu, F.Y. Li, L.Y. Zhu, Anticancer drug release from a mesoporous silica based nanophotocage regulated by either a one- or two-photon process, J. Am. Chem. Soc. 132 (2010) 10645–10647.
[73] X. Cattoën, A. Noureddine, J. Croissant, N. Moitra, K. Bürglová, J. Hodačová, O. Cobos, M. Lejeune, F. Rossignol, D. Toulemon, S. Bégin-Colin, B. Pichon, L. Raehm, J.-O. Durand, M. Wong Chi Man, Click approaches in sol-gel chemistry, J. Sol-Gel Sci. Technol. 70 (2014) 245–253.
[74] S. Dib, M. Boufatit, S. Chelouaou, F. Sadi-Hassaine, J. Croissant, J. Long, L. Raehm, C. Charnay, J.O. Durand, Versatile heavy metals removal via magnetic mesoporous nanocontainers, RSC Adv. 4 (2014) 24838–24841.
[75] R.R. Castillo, A. Baeza, M. Vallet-Regi, Recent applications of the combination of mesoporous silica nanoparticles with nucleic acids: development of bioresponsive devices, carriers and sensors, Biomater. Sci. 5 (2017) 353–377.
[76] J. Zhu, Y. Niu, Y. Li, Y. Gong, H. Shi, Q. Huo, Y. Liu, Q. Xu, Stimuli-responsive delivery vehicles based on mesoporous silica nanoparticles: recent advances and challenges, J. Mater. Chem. B 5 (2017) 1339–1352.
[77] R.R. Castillo, M. Colilla, M. Vallet-Regí, Advances in mesoporous silica-based nanocarriers for co-delivery and combination therapy against cancer, Expert Opin. Drug Deliv. 14 (2017) 229–243.
[78] G.V. Deodhar, M.L. Adams, B.G. Trewyn, Controlled release and intracellular protein delivery from mesoporous silica nanoparticles, Biotechnol. J. 12 (2017) 1600408.
[79] Y. Song, Y. Li, Q. Xu, Z. Liu, Mesoporous silica nanoparticles for stimuli-responsive controlled drug delivery: advances, challenges, and outlook, Int. J. Nanomedicine 12 (2017) 87–110.
[80] D. Niu, Y. Li, J. Shi, Silica/organosilica cross-linked block copolymer micelles: a versatile theranostic platform, Chem. Soc. Rev. 46 (2017) 569–585.
[81] J.L. Paris, M. Colilla, I. Izquierdo-Barba, M. Manzano, M. Vallet-Regí, Tuning mesoporous silica dissolution in physiological environments: a review, J. Mater. Sci. 52 (2017) 8761–8771.
[82] M. Ferre, R. Pleixats, M. Wong Chi Man, X. Cattoen, Recyclable organocatalysts based on hybrid silicas, Green Chem. 18 (2016) 881–922.
[83] Y. Chen, J. Shi, Chemistry of mesoporous organosilica in nanotechnology: molecularly organic-inorganic hybridization into frameworks, Adv. Mater. 28 (2016) 3235–3272.
[84] O. Mongin, L. Porrès, L. Moreaux, J. Mertz, M. Blanchard-Desce, Synthesis and photophysical properties of new conjugated fluorophores designed for two-photon-excited fluorescence, Org. Lett. 4 (2002) 719–722.
[85] J.G. Croissant, O. Mongin, V. Hugues, M. Blanchard-Desce, X. Cattoën, M. Wong Chi Man, V. Stojanovic, C. Charnay, M. Maynadier, M. Gary-Bobo, M. Garcia, L. Raehm, J.-O. Durand, Influence of the synthetic method on the properties of mesoporous silica nanoparticles designed for two-photon excitation, J. Mater. Chem. B 3 (2015) 5182–5188.

[86] M. Pawlicki, H.A. Collins, R.G. Denning, H.L. Anderson, Two-photon absorption and the design of two-photon dyes, Angew. Chem. Int. Ed. 48 (2009) 3244–3266.
[87] J.L. Mynar, A.P. Goodwin, J.A. Cohen, Y. Ma, G.R. Fleming, J.M. Fréchet, Two-photon degradable supramolecular assemblies of linear-dendritic copolymers, Chem. Commun. (2007) 2081–2082.
[88] J. Dong, M. Xue, J.I. Zink, Functioning of nanovalves on polymer coated mesoporous silica nanoparticles, Nanoscale 5 (2013) 10300–10306.
[89] J.-O. Durand, M. Gary-Bobo, P. Maillard, M. Blanchard-Desce, A. Morère, M. Garcia, L. Raehm, Functionalized mesoporous silica nanoparticles for one and two-photon photodynamic therapy, Photodiagn. Photodyn. Ther. 17 (2017) A24.
[90] J.G. Croissant, C. Qi, M. Maynadier, X. Cattoën, M. Wong Chi Man, L. Raehm, O. Mongin, M. Blanchard-Desce, M. Garcia, M. Gary-Bobo, J.-O. Durand, Multifunctional gold-mesoporous silica nanocomposites for enhanced two-photon imaging and therapy of cancer cells. Front. Mol. Biosci. 3 (2016) 1, https://doi.org/10.3389/fmolb.2016.00001.
[91] J. Croissant, M. Maynadier, O. Mongin, V. Hugues, M. Blanchard-Desce, A. Chaix, X. Cattoën, M. Wong Chi Man, A. Gallud, M. Gary-Bobo, M. Garcia, L. Raehm, J.-O. Durand, Enhanced two-photon fluorescence imaging and therapy of cancer cells via gold@bridged silsesquioxane nanoparticles, Small 11 (2015) 295–299.
[92] E. Secret, M. Maynadier, A. Gallud, A. Chaix, E. Bouffard, M. Gary-Bobo, N. Marcotte, O. Mongin, K. El Cheikh, V. Hugues, M. Auffan, C. Frochot, A. Morère, P. Maillard, M. Blanchard-Desce, M.J. Sailor, M. Garcia, J.-O. Durand, F. Cunin, Two-photon excitation of porphyrin-functionalized porous silicon nanoparticles for photodynamic therapy, Adv. Mater. 26 (2014) 7643–7648.
[93] E. Secret, M. Maynadier, A. Gallud, M. Gary-Bobo, A. Chaix, E. Belamie, P. Maillard, M.J. Sailor, M. Garcia, J.-O. Durand, Anionic porphyrin-grafted porous silicon nanoparticles for photodynamic therapy, Chem. Commun. 49 (2013) 4202–4204.
[94] M. Gary-Bobo, Y. Mir, C. Rouxel, D. Brevet, I. Basile, M. Maynadier, O. Vaillant, O. Mongin, M. Blanchard-Desce, A. Morère, M. Garcia, J.-O. Durand, L. Raehm, Mannose-functionalized mesoporous silica nanoparticles for efficient two-photon photodynamic therapy of solid tumors, Angew. Chem. Int. Ed. 123 (2011) 11627–11631.
[95] V. Lebret, L. Raehm, J.O. Durand, M. Smaihi, M.H.V. Werts, M. Blanchard-Desce, D. Methy-Gonnod, C. Dubernet, Folic acid-targeted mesoporous silica nanoparticles for two-photon fluorescence, J. Biomed. Nanotechnol. 6 (2010) 176–180.
[96] C.M. Jimenez, N.Z. Knezevic, Y.G. Rubio, S. Szunerits, R. Boukherroub, F. Teodorescu, J.G. Croissant, O. Hocine, M. Seric, L. Raehm, V. Stojanovic, D. Aggad, M. Maynadier, M. Garcia, M. Gary-Bobo, J.-O. Durand, Nanodiamond-PMO for two-photon PDT and drug delivery, J. Mater. Chem. B 4 (2016) 5803–5808.
[97] C.J. Brinker, G.W. Scherer, Sol-Gel Science: The Physics and Chemistry of Sol-Gel Processing, Academic Press, 2013.
[98] J. Sun, E. Jakobsson, Y. Wang, C. Jeffrey Brinker, Nanoporous silica-based protocells at multiple scales for designs of life and nanomedicine, Life 5 (2015) 214–229.
[99] D. Tarn, C.E. Ashley, M. Xue, E.C. Carnes, J.I. Zink, C.J. Brinker, Mesoporous silica nanoparticle nanocarriers: biofunctionality and biocompatibility, Acc. Chem. Res. 46 (2013) 792–801.
[100] J.L. Townson, Y.-S. Lin, J.O. Agola, E.C. Carnes, H.S. Leong, J.D. Lewis, C.L. Haynes, C.J. Brinker, Re-examining the size/charge paradigm: differing in vivo characteristics of size and charge-matched mesoporous silica nanoparticles, J. Am. Chem. Soc. 135 (2013) 16030.
[101] J. Lu, E. Choi, F. Tamanoi, J.I. Zink, Light-activated nanoimpeller-controlled drug release in cancer cells, Small 4 (2008) 421–426.

[102] J. Moreno, M. Gerecke, L. Grubert, S.A. Kovalenko, S. Hecht, Sensitized two-NIR-photon Z→E isomerization of a visible-light-addressable bistable azobenzene derivative, Angew. Chem. Int. Ed. 55 (2016) 1544–1547.
[103] S. Angelos, E. Johansson, J.F. Stoddart, J.I. Zink, Mesostructured silica supports for functional materials and molecular machines, Adv. Funct. Mater. 17 (2007) 2261–2271.
[104] S. Angelos, E. Choi, F. Voegtle, L. De Cola, J.I. Zink, Photo-driven expulsion of molecules from mesostructured silica nanoparticles, J. Phys. Chem. C 111 (2007) 6589–6592.
[105] D. Tarn, D.P. Ferris, J.C. Barnes, M.W. Ambrogio, J.F. Stoddart, J.I. Zink, A reversible light-operated nanovalve on mesoporous silica nanoparticles, Nanoscale 6 (2014) 3335–3343.
[106] D.P. Ferris, Y.L. Zhao, N.M. Khashab, H.A. Khatib, J.F. Stoddart, J.I. Zink, Light-operated mechanized nanoparticles, J. Am. Chem. Soc. 131 (2009) 1686–1688.
[107] N. Liu, K. Yu, B. Smarsly, D.R. Dunphy, Y.-B. Jiang, C.J. Brinker, Self-directed assembly of photoactive hybrid silicates derived from an azobenzene-bridged silsesquioxane, J. Am. Chem. Soc. 124 (2002) 14540–14541.
[108] J. Britton, E. Antunes, T. Nyokong, Fluorescence quenching and energy transfer in conjugates of quantum dots with zinc and indium tetraamino phthalocyanines, J. Photoch. Photobiol. A 210 (2010) 1–7.
[109] J. Britton, E. Antunes, T. Nyokong, Fluorescence studies of quantum dots and zinc tetraamino phthalocyanine conjugates, Inorg. Chem. Commun. 12 (2009) 828–831.
[110] S.S. Narayanan, S.S. Sinha, S.K. Pal, Sensitized emission from a chemotherapeutic drug conjugated to CdSe/ZnS QDs, J. Phys. Chem. C 112 (2008) 12716–12720.
[111] J.M. Tsay, M. Trzoss, L.X. Shi, X.X. Kong, M. Selke, M.E. Jung, S. Weiss, Singlet oxygen production by peptide-coated quantum dot-photosensitizer conjugates, J. Am. Chem. Soc. 129 (2007) 6865–6871.
[112] X. He, X. Wu, K. Wang, B. Shi, L. Hai, Methylene blue-encapsulated phosphonate-terminated silica nanoparticles for simultaneous in vivo imaging and photodynamic therapy, Biomaterials 30 (2009) 5601–5609.
[113] X. Shen, L. Li, H. Wu, S.Q. Yao, Q.-H. Xu, Photosensitizer-doped conjugated polymer nanoparticles for simultaneous two-photon imaging and two-photon photodynamic therapy in living cells, Nanoscale 3 (2011) 5140–5146.
[114] D. Gao, R.R. Agayan, H. Xu, M.A. Philbert, R. Kopelman, Nanoparticles for two-photon photodynamic therapy in living cells, Nano Lett. 6 (2006) 2383–2386.
[115] D. Kim, J. Kang, T. Wang, H.G. Ryu, J.M. Zuidema, J. Joo, M. Kim, Y. Huh, J. Jung, K.H. Ahn, Two-photon in vivo imaging with porous silicon nanoparticles, Adv. Mater. 29 (2017) 1703309.
[116] N.Ž. Knežević, V. Stojanovic, A. Chaix, E. Bouffard, K. El Cheikh, A. Morère, M. Maynadier, G. Lemercier, M. Garcia, M. Gary-Bobo, Ruthenium (II) complex-photosensitized multifunctionalized porous silicon nanoparticles for two-photon near-infrared light responsive imaging and photodynamic cancer therapy, J. Mater. Chem. B 4 (2016) 1337–1342.
[117] A. Chaix, K. El Cheikh, E. Bouffard, M. Maynadier, D. Aggad, V. Stojanovic, N. Knezevic, M. Garcia, P. Maillard, A. Morère, Mesoporous silicon nanoparticles for targeted two-photon theranostics of prostate cancer, J. Mater. Chem. B 4 (2016) 3639–3642.
[118] H.-L. Tu, Y.-S. Lin, Y. Hung, L.-W. Lo, Y.-F. Chen, C.-Y. Mou, In vitro studies of functionalized mesoporous silica nanoparticles for photodynamic therapy, Adv. Mater. 21 (2009) 172–174.
[119] J. Qian, D. Wang, F. Cai, Q. Zhan, Y. Wang, S. He, Photosensitizer encapsulated organically modified silica nanoparticles for direct two-photon photodynamic therapy and in vivo functional imaging, Biomaterials 33 (2012) 4851–4860.

[120] C. Mauriello-Jimenez, M. Henry, D. Aggad, L. Raehm, C. Charnay, X. Cattoen, M.W.C. Man, S. Alpugan, V. Ahsen, D.K. Tarakcı, Porphyrin-or phthalocyanine-bridged silsesquioxane theranostic nanoparticles for twophoton imaging, photodynamic therapy or photoacoustic imaging, Nanoscale 9 (43) (2017) 16622–16626.
[121] S. Kim, T.Y. Ohulchanskyy, H.E. Pudavar, R.K. Pandey, P.N. Prasad, Organically modified silica nanoparticles co-encapsulating photosensitizing drug and aggregation-enhanced two-photon absorbing fluorescent dye aggregates for two-photon photodynamic therapy, J. Am. Chem. Soc. 129 (2007) 2669–2675.
[122] S.-H. Cheng, C.-C. Hsieh, N.-T. Chen, C.-H. Chu, C.-M. Huang, P.-T. Chou, F.-G. Tseng, C.-S. Yang, C.-Y. Mou, L.-W. Lo, Well-defined mesoporous nanostructure modulates three-dimensional interface energy transfer for two-photon activated photodynamic therapy, Nano Today 6 (2011) 552–563.
[123] H. Liu, Y. Yang, A. Wang, M. Han, W. Cui, J. Li, Hyperbranched polyglycerol-doped mesoporous silica nanoparticles for one-and two-photon activated photodynamic therapy, Adv. Funct. Mater. 26 (2016) 2561–2570.
[124] J.G. Croissant, S. Picard, D. Aggad, M. Klausen, C. Mauriello-Jimenez, M. Maynadier, O. Mongin, G. Clermont, E. Genin, X. Cattoen, M. Wong Chi Man, L. Raehm, M. Garcia, M. Gary-Bobo, M. Blanchard-Desce, J.-O. Durand, Fluorescent periodic mesoporous organosilica nanoparticles dual-functionalized via click chemistry for two-photon photodynamic therapy in cells, J. Mater. Chem. B 4 (2016) 5567–5574.
[125] L. Dykman, N. Khlebtsov, Gold nanoparticles in biomedical applications: recent advances and perspectives, Chem. Soc. Rev. 41 (2012) 2256–2282.
[126] N. Khlebtsov, L. Dykman, Biodistribution and toxicity of engineered gold nanoparticles: a review of in vitro and in vivo studies, Chem. Soc. Rev. 40 (2011) 1647–1671.
[127] S.J. Chadwick, D. Salah, P.M. Livesey, M. Brust, M. Volk, Singlet oxygen generation by laser irradiation of gold nanoparticles, J. Phys. Chem. C 120 (2016) 10647–10657.
[128] M. Yamamoto, I. Osaka, K. Yamashita, H. Hasegawa, R. Arakawa, H. Kawasaki, Effects of ligand species and cluster size of biomolecule-protected Au nanoclusters on efficiency of singlet-oxygen generation, JOL 180 (2016) 315–320.
[129] X. Huang, S. Neretina, M.A. El-Sayed, Gold nanorods: from synthesis and properties to biological and biomedical applications, Adv. Mater. 21 (2009) 4880–4910.
[130] L. Gao, J. Fei, J. Zhao, H. Li, Y. Cui, J. Li, Hypocrellin-loaded gold nanocages with high two-photon efficiency for photothermal/photodynamic cancer therapy in vitro, ACS Nano 6 (2012) 8030–8040.
[131] C. Jiang, T. Zhao, P. Yuan, N. Gao, Y. Pan, Z. Guan, N. Zhou, Q.-H. Xu, Two-photon induced photoluminescence and singlet oxygen generation from aggregated gold nanoparticles, ACS Appl. Mater. Interfaces 5 (2013) 4972–4977.
[132] P. Zijlstra, J.W.M. Chon, M. Gu, Five-dimensional optical recording mediated by surface plasmons in gold nanorods, Nature 459 (2009) 410–413.
[133] N.-T. Chen, K.-C. Tang, M.-F. Chung, S.-H. Cheng, C.-M. Huang, C.-H. Chu, P.-T. Chou, J.S. Souris, C.-T. Chen, C.-Y. Mou, L.-W. Lo, Enhanced plasmonic resonance energy transfer in mesoporous silica-encased gold nanorod for two-photon-activated photodynamic therapy, Theranostics 4 (2014) 798–807.
[134] B. Moosa, K. Fhayli, S. Li, K. Julfakyan, A. Ezzeddine, N.M. Khashab, Applications of nanodiamonds in drug delivery and catalysis, J. Nanosci. Nanotechnol. 14 (2014) 332–343.
[135] M. Gary-Bobo, O. Hocine, D. Brevet, M. Maynadier, L. Raehm, S. Richeter, V. Charasson, B. Loock, A. Morere, P. Maillard, M. Garcia, J.-O. Durand, Cancer therapy improvement with mesoporous silica nanoparticles combining targeting, drug delivery and PDT, Int. J. Pharm. 423 (2012) 509–515.
[136] J. Croissant, Y. Fatieiev, K. Julfakyan, J. Lu, A. Emwas, D. Anjum, H. Omar, F. Tamanoi, J. Zink, N. Khashab, Biodegradable oxamide-phenylene-based

mesoporous organosilica nanoparticles with unprecedented drug payloads for delivery in cells, Chem. A Eur. J. 22 (2016) 14806–14811.
[137] J.G. Croissant, Y. Fatieiev, H. Omar, D.H. Anjum, A. Gurinov, J. Lu, F. Tamanoi, J.I. Zink, N.M. Khashab, Periodic mesoporous organosilica nanoparticles with controlled morphologies and high drug/dye loadings for multicargo delivery in cancer cells, Chem. A Eur. J. 22 (2016) 9607–9615.
[138] J.G. Croissant, X. Cattoën, J.-O. Durand, M.W.C. Man, N.M. Khashab, Organosilica hybrid nanomaterials with a high organic content: syntheses and applications of silsesquioxanes, Nanoscale 8 (2016) 19945–19972.
[139] Y. Fatieiev, J. Croissant, K. Julfakyan, L. Deng, D.H. Anjum, A. Gurinov, N. Khashab, Enzymatically degradable hybrid organic–inorganic bridged silsesquioxane nanoparticles for in vitro imaging, Nanoscale 7 (2015) 15046–15050.
[140] S.A. Shahin, B.T. Vu, J.G. Croissant, Y. Fatieviev, K. Matsumoto, T.L. Hoang Doan, T. Yik, A. Conteras, L. Ratliff, C. Mauriello Jimenez, L. Raehm, N.M. Khashab, J.-O. Durand, C. Glackin, F. Tamanoi, Chick chorioallantoic membrane assay as an in vivo model to study the effect of nanoparticle-based anticancer drugs in ovarian cancer, Sci. Rep. 8 (2018) 8524.
[141] J. Croissant, D. Salles, M. Maynadier, O. Mongin, V. Hugues, M. Blanchard-Desce, X. Cattoën, M. Wong Chi Man, A. Gallud, M. Garcia, M. Gary-Bobo, L. Raehm, J.-O. Durand, Mixed periodic mesoporous organosilica nanoparticles and core-shell systems, application to in vitro two-photon imaging, therapy and drug delivery, Chem. Mater. 26 (2014) 7214–7220.
[142] J.G. Croissant, X. Cattoën, M. Wong Chi Man, J.O. Durand, N.M. Khashab, Syntheses and applications of periodic mesoporous organosilica nanoparticles, Nanoscale 7 (2015) 20318–20334.
[143] W. Huybrechts, G. Mali, P. Kuśtrowski, T. Willhammar, M. Mertens, S. Bals, P. Van Der Voort, P. Cool, Post-synthesis bromination of benzene bridged PMO as a way to create a high potential hybrid material, Microporous Mesoporous Mater. 236 (2016) 244–249.
[144] M.D. Esquivel Merino, E. De Canck, C. Jiménez-Sanchidrián, P. Van Der Voort, F.J. Romero-Salguero, Periodic mesoporous organosilicas as catalysts for organic reactions, Curr. Org. Chem. 18 (2014) 1280–1295.
[145] P. Der Voort, Ultra-fast hydrothermal synthesis of diastereoselective pure ethenylene-bridged periodic mesoporous organosilicas, Chem. Commun. (2007) 2261–2263.
[146] J. Croissant, X. Cattoën, M.W.C. Man, A. Gallud, L. Raehm, P. Trens, M. Maynadier, J.O. Durand, Biodegradable ethylene-bis (propyl) disulfide-based periodic mesoporous organosilica nanorods and nanospheres for efficient in-vitro drug delivery, Adv. Mater. 26 (2014) 6174–6180.
[147] J.G. Croissant, Y. Fatieiev, N.M. Khashab, Degradability and clearance of silicon, organosilica, silsesquioxane, silica mixed oxide, and mesoporous silica nanoparticles, Adv. Mater. 29 (2017) 1604634.
[148] R. Bouchal, M. Daurat, M. Gary-Bobo, A. Da Silva, L. Lesaffre, D. Aggad, A. Godefroy, P. Dieudonné, C. Charnay, J.-O. Durand, Biocompatible periodic mesoporous ionosilica nanoparticles with ammonium walls: application to drug delivery, ACS Appl. Mater. Interfaces 9 (2017) 32018–32025.
[149] S. Datz, H. Engelke, C.V. Schirnding, L. Nguyen, T. Bein, Lipid bilayer-coated curcumin-based mesoporous organosilica nanoparticles for cellular delivery, Microporous Mesoporous Mater. 225 (2016) 371–377.
[150] S. Biswas, H.-Y. Ahn, M.V. Bondar, K.D. Belfield, Two-photon absorption enhancement of polymer-templated porphyrin-based J-aggregates, Langmuir 28 (2011) 1515–1522.
[151] R. Won, Endoscopy: two-photon imaging, Nat. Photonics 10 (2016) 73.

CHAPTER FIVE

Controlled Release With Emphasis on Ultrasound-Induced Release

Maria Vallet-Regi[*,†,1], **Miguel Manzano**[*,†], **Alejandro Baeza**[*,†,‡]
[*]Dpto. Química Inorgánica y Bioinorgánica, Universidad Complutense de Madrid, Instituto de Investigación Sanitaria Hospital 12 de Octubre i+12
[†]CIBER de Bioingeniería, Biomateriales y Nanomedicina, CIBER-BBN, Madrid, Spain
[‡]Dpto. Materiales y Producción Aeroespacial, ETSI Aeronáutica y del Espacio, Universidad Politécnica de Madrid, Madrid, Spain
[1]Corresponding author: e-mail address: vallet@ucm.es

Contents

Abstract

Mesoporous silica nanoparticles (MSNs) represent a promising approach to be used as nanocarriers because they fulfill some basic criteria: biocompatibility, high loading capacity and possibility of easy surface modifications. As a consequence, MSNs have been employed to design sophisticated stimuli-responsive nanocarriers able to release the entrapped cargo on demand. Among those stimuli ultrasound (US) is considered as one of the most promising triggers for drug delivery systems due to the high tissue-penetrating capability and non-invasiveness. Recently, we employed US as a stimulus to control the drug release from MSNs for the first time, and found that acoustic cavitation played a key role in triggering the release. Those US-responsive MSNs were transported using cellular vehicles toward disease areas where they demonstrated their capacity to kill cancer cells when exposed to US.

Light is also an excellent trigger because of the high biocompatibility and the capacity to focus to very small regions. However, its poor penetration in living tissues compromises its use in a clinical scenario. An alternative could be the use of near infrared (NIR) as trigger, as it would be described in this chapter. Other option could be the use of magnetic fields as stimulus because it can reach deep areas into the living tissue

The Enzymes, Volume 43
ISSN 1874-6047
https://doi.org/10.1016/bs.enz.2018.07.005

without damaging the surrounding tissues. MSNs have been combined with magnetic fields in a many different approaches and a quick overview is here provided.

This chapter will show how some of these external stimuli represent a valuable tool for fighting complex diseases such as cancer.

1. INTRODUCTION

In the last few years nanoparticles able to host, protect, transport and release therapeutic agents to specific tissues have been deeply investigated for biomedical applications. Among them, mesoporous silica nanoparticles (MSNs) fulfill some of the requirements that nanoparticles should present to be used as drug delivery systems: high loading capacity and many different surface modifications possibilities. The reason for the increased interest on MSNs lies on their excellent properties, such as large surface areas (ca. $1000\,m^2/g$), great pore volumes (ca. $1\,cm^3/g$), tunable pore diameters (2–6 nm), robustness and easy to functionalize. Additionally, the biocompatibility of MSN has been evaluated *in vivo* through many different studies, attending to their biodistribution, toxicity and excretion; those investigations have revealed that MSNs exhibits remarkable good biocompatibility. In fact, there is a phase I/II trial (NCT02106598) that involves targeted silica nanoparticles for image-guided operative sentinel lymph mode mapping.

Since 2001, when our research group proposed MCM-41 mesoporous materials as drug delivery systems for the first time [1], an impressive number of publications regarding the use of MSNs as drug carriers have been observed. The network of cavities allows loading different therapeutic compounds from different nature, from small drug molecules to proteins, among others [2]. However, when the cargo might present certain toxicity levels, as it might be the case of some antitumor drugs, it is important to avoid the premature release of those drugs before reaching the targeted tissues. Then, once the nanocarriers might be at the diseased tissue, the cargo should be released in a controlled fashion. To achieve that, those nanocarriers can be designed placing different moieties on the pore entrances so they would act as gatekeepers hampering the release of the cargo. It is obvious that those gatekeepers should be reversible, that is, they should allow the release of the cargo on demand once the nanocarriers might be at the right place, in what is called stimuli-responsive drug delivery systems [3].

The different stimuli that are commonly employed for triggering the release of the cargo from a nanocarrier can be divided in internal and external stimuli. Internal stimuli are characteristic of the disease to be treated. In the case of tumors, the stimulus is based on the differences between tumor cells and healthy cells. Thus, once the nanoparticles might be at the tumor, the presence of such a stimulus will promote some changes on the nanocarriers that would provoke the drug release. Some typical examples are pH, since tumors present lower pH than healthy tissues, redox potential or the presence of certain proteins or signaling molecules, such a s certain matrix metalloproteases. On the other hand, external stimuli are not a consequence of the pathology to be treated. They are exogenous agents that might be applied by the clinician to the diseased site after the administration of the nanocarriers, and some examples are: light, magnetic field or ultrasound.

2. ULTRASOUND-INDUCED RELEASE

Ultrasound (US) has been widely employed in the clinic both for diagnostic and/or treatment [4, 5]. The reason for that interest relays on its capacity to non-invasively penetrate deep into internal tissues being well tolerated by the body [3]. Among other applications, US has been used for tumor ablation [6], for the treatment of Alzheimer's disease [7], to generate hyperthermia [8], to cross the blood brain barrier [9], for immunostimulation [9], for biomedical imaging [9] and physical therapy [10].

2.1 General Aspects of Ultrasound

Ultrasound is normally generated from piezoelectric transducers that are able to transform an electric signal to a mechanic wave that can travel along a fluid. It is currently being employed in different areas, such as object detection and distance measurement, imaging in medicine and different chemical processes in industry. Ultrasound can be divided into low (<1 MHz), medium (1–5 MHz) and high (5–10 MHz) frequency US. Low frequency US present better penetration depth in tissues and are non-invasive, but in contrast they cannot be focused into a spot. On the contrary, US with frequencies higher than 2 MHz present poor tissue penetration and due to the increased scattering they might damage the tissue due to the produced heating. However, they can be focused and the intensity at the focal point becomes higher than away from that point. Curved transducers allow to focus high intensity US inside the body, the so-called high intensity focused ultrasound (HIFU), allowing the application of the stimulus for certain therapies such as the destruction of internal tumor tissue from the outside.

The biological effects of US can be divided into two categories: thermal effects and mechanical effects [8]. Thermal effects are produced when the US wave travels across a tissue, part of the energy might be adsorbed by the tissue causing a rise of temperature. Thus, that hyperthermia generated by the US could be employed in ablative therapies against cancer. These thermal effects are directly related to the US frequency, in a way that the higher is the later, the greater is the temperature rise when traveling across tissues.

Among mechanical effects, cavitation is the most useful phenomena employed in biomedicine, and it can be defined as the formation and oscillation of gas bubbles in a fluid [11]. Those bubbles, which are formed as a consequence of the exposition of the fluid to the mechanical waves from US, can oscillate about an equilibrium bubble radius for several acoustic cycles. However, at certain ultrasound frequency and bubble size, and working at elevated pressures, the bubble can grow unstable and collapse violently during compression. This phenomenon, known as inertial cavitation, produces extreme conditions in the local environment in terms of high temperature and pressure at the nanoscale level. This is the reason why both effects normally go together and it is very difficult to separate them. Those extreme conditions lead to the emission of short bursts of light (sonoluminescence) and reactive oxygen species coming from the pyrolysis of the water molecules.

Thermal and mechanical effects can be employed to induce or accelerate chemical reactions, in what is called sonochemistry [12]. The induction of chemical reactions in response to inertial cavitation can be explained by the generation of reactive oxygen species that lead to subsequent chemical reactions, by the heating at the nanoscale level in the local environment of the bubbles implosion, by some mechanical effects such as acoustic microflow associated to cavitation or by a combination of all of them.

2.2 US-Responsive Materials

The thermal, mechanical and chemical effects from US have been employed to develop different smart materials able to respond to the application of the stimulus. In fact, US characteristics (high tissue-penetrating capability, non-invasiveness, portability and spatiotemporal controllability) has fueled its application in biomedicine, being generally regarded as one of the most promising options for controlled drug delivery [13, 14, 15].

2.3 US-Responsive MSNs

In the last few years, there has been an increased interest on developing smart mesoporous nanoparticles able to release their cargo in response to certain stimuli. Mesoporous silica nanoparticles present an open structure, so it is necessary to decorate them with a moiety to close the pore entrances and avoid premature release. Additionally, that moiety should be able to allow the release of the cargo on demand. The nature of the materials controlling the release behavior from MSNs has been widely explored, ranging from polymers to small molecules acting as pore gates.

Among the available stimuli for controlling the cargo release, ultrasound seems to be very interesting to be used in biomedicine, as it has been described in this chapter, because of the high tissue-penetrating capability, non-invasiveness, portability and spatiotemporal controllability. In this sense, the combination of mesoporous silica materials with US for biomedical applications has been centered on bulk materials. An example is capping the surface of mesoporous silica with poly(dimethylsiloxane), which responded to US irradiation promoting the release of ibuprofen [16].

It has been only in the last few of years when US-triggered stimuli-responsive drug delivery systems based on MSNs have been developed [17, 18, 19]. The reason for such as delay in comparison with other type of stimuli such as light triggering or endogenous pH relays on the difficulty to design and/or produce US-responsive nanovalves on the surface of mesoporous materials.

Our research group has recently developed a nanocarrier based on MSNs decorated with a *co*-polymer that presents a US-cleavable moiety [20]. The employed *co*-polymer, p(MEO_2MA)-*co*-THPMA, p(MEO_2MA): poly (2-(2-methoxyethoxy)ethyl methacrylate and THPMA: 2-tetrahydropyranyl methacrylate), presents a labile acetal group that can be cleaved by US to yield a different polymer with different hydrophobicity. Thus, we took advantage of the phase transformation from hydrophobic to hydrophilic by US to design a polymeric gatekeeper on MSNs to release the cargo under control. The selected *co*-polymer shows a low critical solution temperature (LCST) bellow 37°C, so at 4°C the *co*-polymer is in a coil-like conformation, which allows the loading of the cargo into the mesopores. Then, when the nanocarriers are exposed to physiological temperature, the *co*-polymer changes its conformation toward a collapsed state, due to the insolubility in aqueous medium, which closes the pores entrances retaining the cargo inside the mesopores. This is an important feature of our system, since it is possible to produce the nanocarriers with the gates already grafted and then, in a separate step, it is possible to load the desired drug in a simple way: locate the nanocarriers solution in a fridge, add the drug solution to the nanoparticles and place the loaded nanoparticles at 37°C so the polymer conformation will change closing the pore entrances with the cargo already loaded. When US might be irradiated to the loaded nanocarriers, the hydrophobic moiety from the THPMA is cleaved so the hydrophilicity of the *co*-polymer increases and, consequently, the LCST also increases over physiological temperature. As a consequence, a change in the conformation of the polymer takes place, from collapsed to extended, which opens the gates of the mesopores and leads to the release of the entrapped cargo (Fig. 1).

The biocompatibility of designed US-responsive nanocarriers was evaluated *in vitro* incubating them with tumor cells (LNCaP cells, from human prostate adenocarcinoma). None of the byproducts from the nanoparticles before and after US exposure were observed to affect the cell viability, ensuring the lack of toxicity of the designed system. The lack of damage to relevant biological molecules produced by the US exposure was also ensured using a commercial US apparatus commonly employed for physical therapy in humans working at 1 MHz and 15 W.

The intracellular release of the cargo was evaluated labeling the MSNs with rhodamine-B and loading the mesopores of the US-responsive system with fluorescein. As it can be observed in Fig. 2, before US exposure both green and red from fluorescein and rhodamine, respectively, *co*-localized, suggesting that fluorescein was entrapped inside the nanocarriers. Then, after 5 min of US irradiation, the cell cytoplasm was stained in green because

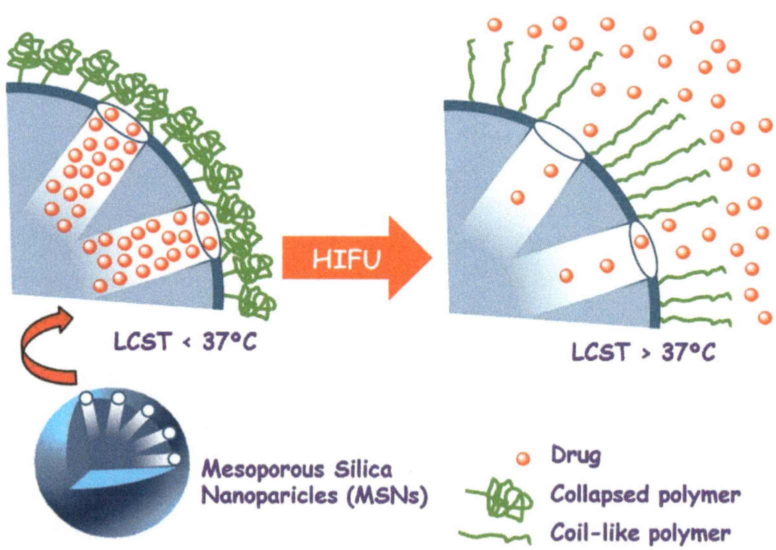

Fig. 1 Schematic representation of the cargo release mechanism of US-responsive MSNs when exposed to high intensity focused ultrasound (HIFU).

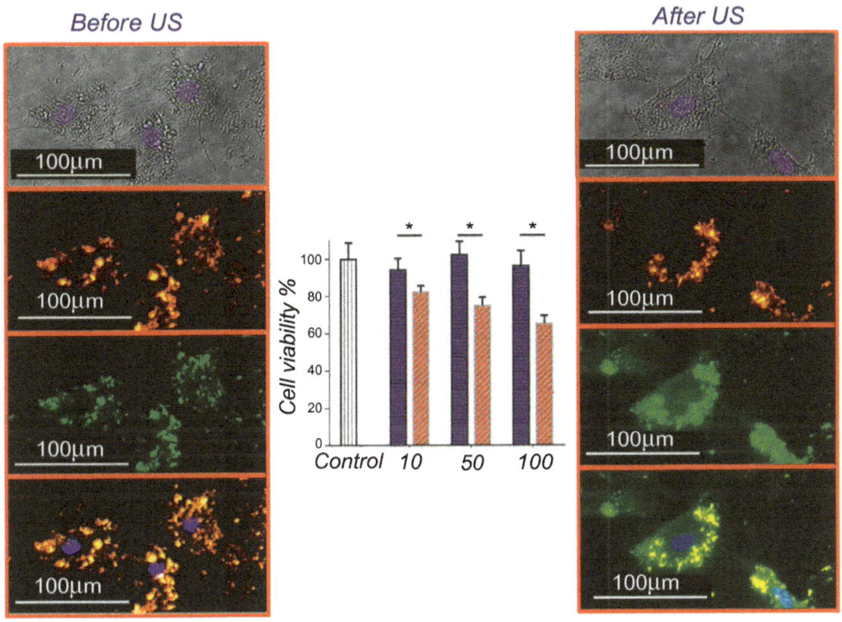

Fig. 2 (*Left*) Fluorescence microscopy images of cancer cells incubated with rhodamine B-labeled US-sensitive MSNs loaded with fluorescein before US irradiation. (*Center*) Cell viability experiment in LNCaP cells with different concentrations of doxorubicin loaded US-sensitive MSNs. (*Right*) Fluorescence microscopy images of cancer cells incubated with rhodamine B-labeled US-sensitive MSNs loaded with fluorescein after US irradiation.

of the fluorescein release from the carriers, which meant that the nanogates of the pores were opened as a consequence of the US, proving that the designed nanocarriers retained the US-responsive capability after cellular internalization. Taking into account the potential application of this technology for anticancer therapy, the nanocarriers were loaded with doxorubicin and evaluated with cancer cells. While no toxicity was observed before US application, which meant that there was not premature release, a significant decrease in cancer cells viability after US exposition (Fig. 2), which meant that the nanocarriers were responding to the US and releasing the cytotoxic drug to kill the cancer cells.

One of the main challenges in nanomedicine for the potential treatment of cancer relays on the selective accumulation of the nanoparticles into the tumor areas without affecting healthy tissues. Nanoparticles can be targeted and delivered into tumors through either passive targeting, thanks to the enhanced permeability and retention (EPR) effect [21], or active targeting, which consists on the functionalization of the surface of the nanoparticles with some affinity ligands that would be specifically recognized by specific cell receptors of diseased cells [22]. However, those approaches have not achieved the expected results in nanomedicines targeting [23, 24]. This is why different alternatives to localize the nanoparticles in the tumors have been explored. In this sense, the use of cells that migrate toward tumors to transport the nanoparticles has been investigated in the last few years [25]. This is the case of human mesenchymal stem cells, which present inherent migratory properties in response to inflammation or injury [26, 27, 28]. The common sources of adult mesenchymal stem cells are bone marrow and adipose tissue, although they present some drawbacks, such as the invasiveness and inefficient isolation techniques in terms of isolated cell quantities together with the dependence to the donor age. An additional source of mesenchymal stem cells is the human decidua of the human placenta, where decidua mesenchymal stem cells (DMSCs) can be found. Those cells present several advantages over conventional mesenchymal stem cells, such as: easy and non-invasive isolation techniques, they represent a homogenous population of adult stem cells with low or non-immune response and, more importantly, DMSCs exhibit migratory properties toward tumors [29, 30]. Thus, DMSCs seem to be excellent candidates to transport nanomedicines to tumor tissues. In this sense, we have employed those cells to transport our US-responsive MSNs nanoparticles, so the release of the cytotoxic cargo would take place only at the targeted tissue [31]. Thus, the MSNs provide control on the release while the DMSCs provide the specific targeting toward tumors. The previously developed US-responsive MSNs needed to be coated with poly(ethyleneimine) to internalize them into DMSCs favoring the electrostatic interaction between the positively charged polymer and the negatively charged cell membrane. The nanoparticles were non-toxic at conventional concentration (i.e., 200 μg/mL) and they were retained inside the DMSCs for a minimum of 6 days, long enough for those cells to reach the tumor tissue, as it was found in previous work [32]. As it can be observed in Fig. 3, the ultrasound responsiveness of the system was evaluated *in vial*, through a fluorescein release experiment; *in vitro*, after internalization of the rhodamine B-labeled nanoparticles loaded with fluorescein; and *in vivo*, through the subcutaneous injection of calcein-AM loaded MSNs responsive to US into mice, and after US application, the calcein-AM (low-fluorescent indicator) was released and converted to calcein (highly fluorescent indicator) as a consequence of the activity of the estereases of the surrounding cells.

The US-responsive MSNs loaded DMSCs platform were observed to reach the tumor, as observed in an *in vitro* cell migration assay; and when

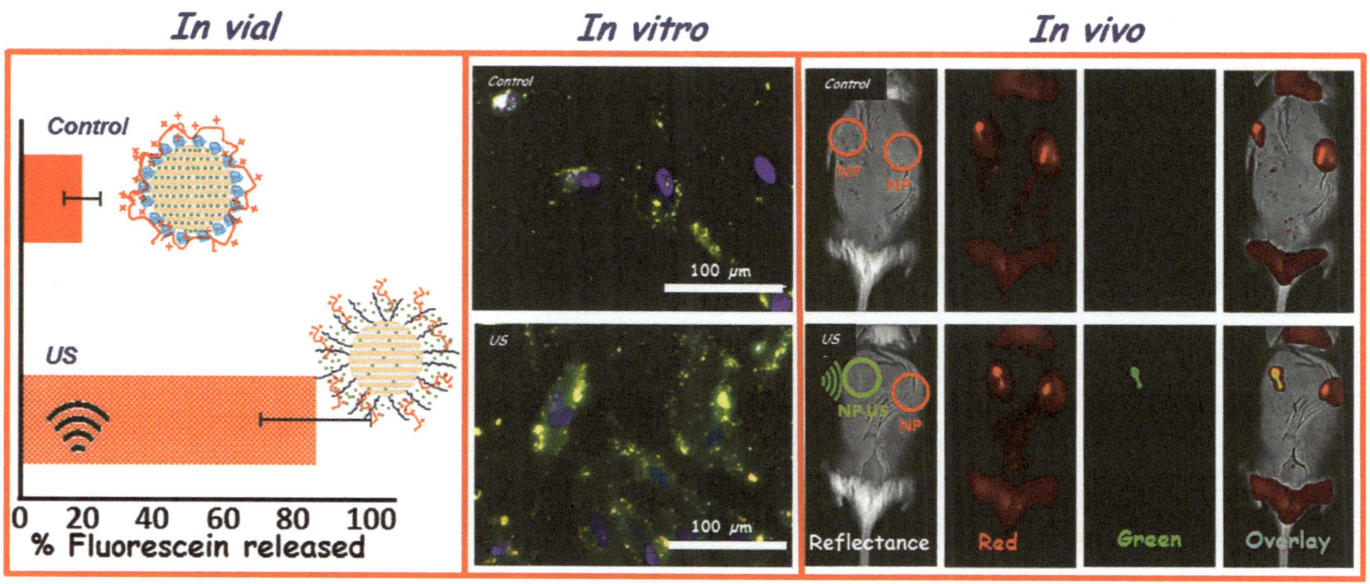

Fig. 3 Different scenarios of PEI-modified US-sensitive MSNs release: *in vial* (*left*), *in vitro* (*center*) and *in vivo* (*right*).

loading the MSNs with doxorubicin, those that were previously exposed to US were observed to reduce NMU cancer cells viability, showing that it is possible to introduce cytotoxic loaded US-responsive MSNs into DMSCs as cell carriers. Those cells kept their migratory properties, so they would transport the loaded nanoparticles to the tumor tissue, and once in there, the platform can be triggered with US, releasing the cytotoxic drug and affecting only at the diseased tissue.

Other authors have recently developed a US-triggered drug release system based on the specific interaction between a mesoporous organosilica framework and drug molecules [33]. The non-covalent interactions between the matrix and the drug molecules, which can be of different nature, such as supramolecular or hydrophobic–hydrophobic interactions, can be easily interrupted by the application of external US. An example of this approach was developed using hollow periodic mesoporous organosilicas with phenyl-bridged framework and loaded with paclitaxel [34].

Hollow periodic mesoporous organosilicas with disulfide bonds within the framework have also been developed as drug carriers that could be broken into pieces and release their cargo as a consequence of the high thermal and mechanical effects from high intensity focused ultrasound [18].

3. MESOPOROUS SILICA NANOPARTICLES WITH LIGHT-RESPONSIVE BEHAVIOR

Light has been widely employed for drug delivery applications due to its excellent properties such the high biocompatibility and the capacity to precisely focus its application in very small regions. Thanks to this last property, a nanodevice engineered to release the housed drug in response to light is able to treat tumoral lesions with excellent precision without affecting the unexposed neighbor tissue. The main problem of light is its poor penetration in living tissues. Ultraviolet and visible light are completely adsorbed in only a few millimeters which compromise their usefulness in the clinical field. Despite this limitation, these light sources can be applied for the treatment of malignant lesions in exposed tissues such as skin, colon, eye or stomach, among others. Moreover, it is even possible to deliver these types of light to internal tumors employing optic fibers by endoscopic techniques. Fujiwara et al. reported in 2003 the use of UV light for triggering the drug departure in MSN [35]. In this case, coumarin molecules were grafted on the pore outlets in order to act as light-responsive gatekeepers. This system was able to retain a hydrophobic cargo (cholestane) when it was exposed to UV light at 324 nm of wavelength as a consequence of the dimerization of the coumarins attached on the pore entrance, whereas it exhibited drug departure when the system was irradiated with a more energetic radiation at 250 nm due to the rupture of these dimers. In a later work, the same research group implemented this system through the incorporation of azobenzene moieties attached on the pore walls [36]. Azobenzene suffers trans-to-cis photoisomerization in response to UV/vis light acting as "nanoimpellers" which propels the housed drugs outside the MSN. This pioneering work paved the way for using light as trigger stimulus of drug release in MSN. Thus, different pore-blockers such as small molecules [37], metal-complexes [38] or polymers [39] have been attached inside the pore walls through light-sensitive bonds in order to retain the housed drugs until the light is applied. In these cases, the employed pore-blockers acted only as caps, but one interesting strategy consists in the use of gate keepers capable to play an additional role, for instance acting as targeting moieties at the same time that close the pore entrance. Martinez-Carmona et al. have reported the use of transferrin attached through a light-cleavable linker in order to trigger the drug release under UV radiation and act as targeting group at the same time [40]. Transferrin is the protein on charge of the transportation of iron in many organism including humans. Due to the accelerated growing rate of tumoral cells, they require large amounts of iron and therefore, they usually overexpress transferrin receptors on their membrane surface. In this work, MSN surface was decorated with a designed linker which contained a biotin molecule at the end and a light-sensitive nitro-benzene moiety in the middle. The system was loaded with a potent cytotoxic compound, doxorubicin (Dox) and then, it was sealed adding streptavidin and biotinylated transferrin thanks to the formation of a strong biotin-streptavidin bridge between the MSN surface and the protein. When this system was exposed to UV light at 366 nm of wavelength, the linker was broken and the transferrin moiety was detached from the surface allowing the drug departure (Fig. 4). The capacity of these nanocarriers to be engulfed by tumoral cells was evaluated in fibrosarcoma cells (HT1080) which overexpress transferrin receptors on their surface, observing an enhanced cellular uptake up to eight times higher when the MSN were decorated with transferrin in comparison with the naked particles. The exposition to UV light caused a potent cytotoxic effect inducing the apoptotic state in the tumoral cells at dosages of nanocarriers lower than 1 $\mu g mL^{-1}$, probably due to the synergy between the cytotoxic effect of the released Dox and the own effect of the UV light in the tumoral

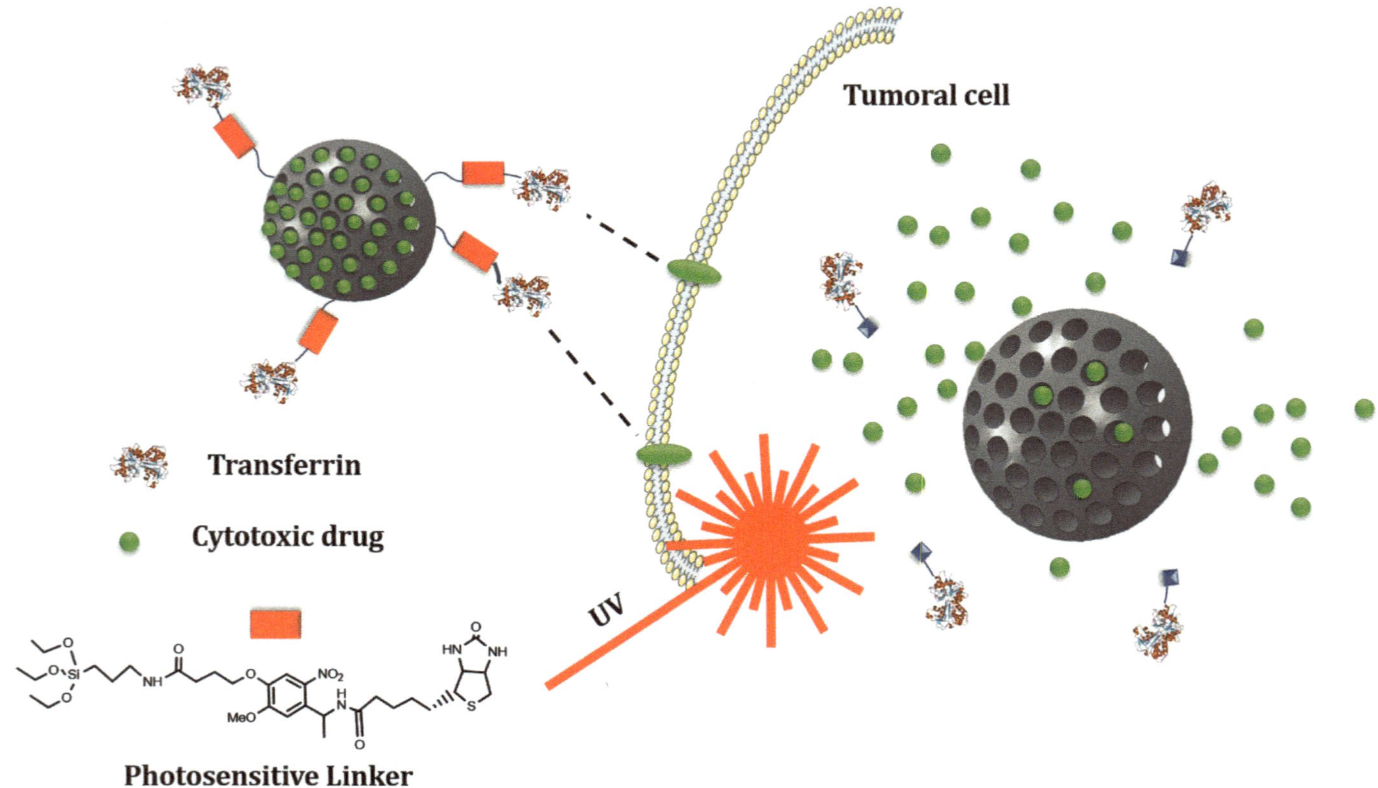

Fig. 4 MSN decorated with transferrin as targeting and light-sensitive cap.

cells. The excellent performance regarding enhanced cellular uptake, synergism and controlled release provided by this nanodevice would allow the treatment of tumoral lesions with high precision reducing the side effects of the chemotherapy.

As it was mentioned earlier, UV light present poor tissue penetration and additionally, it induces cellular damage at prolonged exposition. For this reason, the use of visible light instead of UV represents an excellent alternative due to the harmless nature of this radiation. Hollow MSN were coated with a photodegradable amphiphilic *co*-polymer which retained the drug trapped within the silica core [41]. The polymeric coat was degraded under the exposition to green light (500–540 nm) allowing the drug release. Additionally, the external surface of these core-shell particles was decorated with folic acid, which acted as targeting agent due to the overexpression of folate receptors by many tumoral cell lines. The presence of this targeting group allowed the selective internalization of the nanocarrier in the tumoral cells and the light-responsive behavior which provoked their efficient destruction. Vallet-Regí et al. have reported the use of porphyrin caps attached on the MSN surface by a linker which can be cleavable by the action of radical oxidative species (ROS) [42]. Porphyrins present a similar size to the pore entrances and are able to close the pores retaining topotecan inside the particles. Under exposition to visible light, the porphyrins produced ROS which induced the linker rupture releasing the caps and allowing the topotecan release inside tumoral cells. One of the most important properties of this nanodevice is its ability to escape from endosomes. Practically all nanocarriers enter within the cells *via* endocytosis [43]. Thus, the nanoparticles are engulfed into endosomes/lysosomes and their cargo can be degraded by the acidic environment and digestive enzymes present in these organelles. Endosomal escape is an important ability which should be provided to nanocarriers, especially when they transport sensitive molecules such as proteins or oligonucleotide chains. In this work, the generation of ROS induces the rupture of the endosomes releasing the particles into the cytosol which allow the efficient transportation of therapeutic agents into the diseased cells.

In the recent years, the use of near infrared radiation (NIR) has received huge attention because this light is less adsorbed by tissues and therefore, its penetration is significantly higher than UV or visible light. Single-walled nanotubes have been decorated with Dox-loaded nanoparticles, and the drug departure was achieved exposing this system under NIR at 808 nm of wavelength [44]. The induced Dox release is caused by the heat generated by the interaction of NIR with the carbon surface. Moreover, the temperature increase produces a synergic enhancement of the cytotoxic effect of Dox. The performance of this system was evaluated *in vivo* in a mouse model of breast cancer 4T1 showing that the injection of these nanodevices rapidly increased the temperature in the tumoral zone at 48°C when NIR was applied during 20 min. The temperature increase in combination with the released Dox caused the tumor shrinkage without affecting the rest of the organs. Gold nanorods (GNR) also exhibit strong interaction with NIR due to the plasmon band on the gold surface [45]. Thus, GNR have been coated with a mesoporous silica shell (GNR@MSN) and the pore outlets were capped with sulfonatocalix[4]arene as caps [46]. NIR exposition produced the detachment of calixarenes by the reduction of the stalk binding affinity opening the pores and therefore, releasing the drug trapped inside. One important characteristic of this system is that it is reversible. When the light was switched off, the caps were conjugated again on the MSN surface sealing the pores resulting into a "On–Off" NIR-sensitive nanodevice. GNR@MSN have also been coated with thermosensitive polymers which acted as a diffusion barrier that hampers the premature release of drugs loaded into the porous silica network [47]. The thermosensitive polymer based on poly-*N*-isopropylamide (PNIPAM) was designed to show a transition temperature at 43°C in which the polymer suffers linear-to-globular transition that allows the drug release. In order to provide selectivity to the nanodevice for tumoral cells, in this case melanoma cells, the polypeptide NAPamide was attached on the polymeric surface due to the strong interaction with the alpha melanocyte receptor overexpressed on the surface of these malignant cells. This system was able to destroy the melanoma cells at dosages of 10 $\mu g\,mL^{-1}$ under the exposition to NIR laser at 808 nm during 10 min, while it barely altered the cell viability of healthy fibroblasts used as control (Fig. 5).

4. MESOPOROUS SILICA NANOPARTICLES WITH MAGNETIC-RESPONSIVE DRUG RELEASE

The application of magnetic field for triggering the drug departure has been received huge attention the recent years due to the, in general, excellent biocompatibility of this stimulus. As it was mentioned above, one of the main limitations of light is its poor penetration capacity in living tissues. On the contrary, magnetic fields are able to reach deep zones within the tissue without producing side effects in the zone. MSN are completely inert

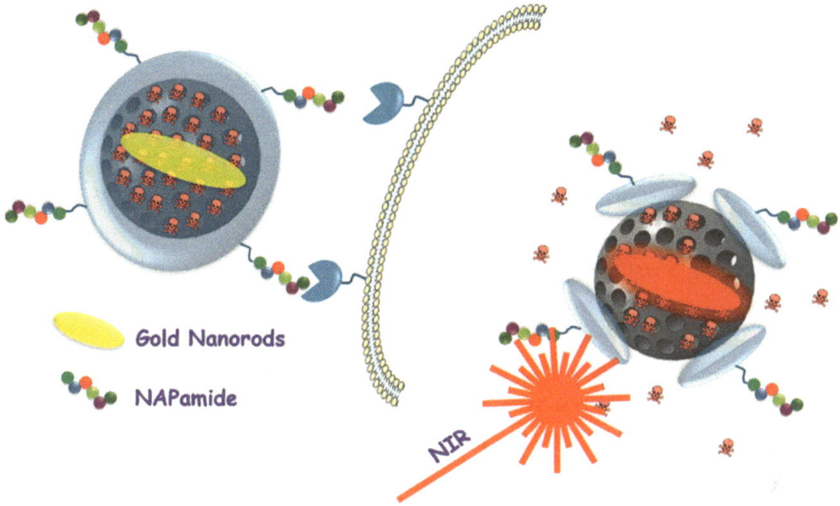

Fig. 5 MSN coated with thermosensitive polymer and NAPamide for NIR-induced drug release.

to magnetic fields and therefore, the use of this stimulus as trigger of the drug departure process require the introduction of a magnetic sensitive core within the silica matrix. One of the most employed systems with this aim has been the utilization of superparamagnetic iron oxide nanoparticles (SPION). SPION interact with alternative magnetic field (AMF) generating heat mainly due to two mechanisms, Brown and Néel relaxation, which are, respectively, based on the heat generated by the friction between the spinning particle and the dispersant and the rotation of the magnetic moment in the particle. SIPON are embedded within the silica matrix forming core-shell magnetic nanoparticles (MSN@SPION) which maintain the porosity of the silica particles and the magnetic properties of the iron oxide. In order to exploit the generated heat for triggering the drug departure, a thermosensitive moiety should be incorporated on the pore outlets which can open or close the system in response to temperature increments. One possibility is to anchor caps that block the pore outlets through temperature-cleavable bond such as the system reported by Kasko et al. in which a polyethylenglycol chains were grafted on the MSN surface employing azo-bonds [48]. Bioorganic moieties as DNA have also been employed for this purpose. Thus, Vallet-Regí et al. have attached single-DNA strands of 15 base pairs on the surface of MSN which contained SPION embedded inside [49]. The system was loaded with fluorescein, as a drug model, and the addition of SPION decorated with the complementary DNA strand sealed the system by the hybridization between both DNA strands. The temperature increase achieved under the exposition to AMF produced the dehybridization of both strands opening the system and allowing the drug departure. When the AMF was switched off, both strands were hybridized again closing the pores yielding a system able to release the therapeutic cargo on demand. Baeza et al. have described the use of a thermosensitive polymer which was grafted around the MSN@SPION surface based on PNIPAM grafted with polyethyleneimine [50]. In this work, the thermosensitive shell was designed with a dual purpose, on the one hand the PNIPAM layer avoided the premature release of drugs trapped within the silica pores and on the other hand, PEI shell retained proteins by electrostatic interactions. This system was able to transport and release in response to AMF exposition therapeutic compounds which presented very different nature, small molecules inside the silica and big macromolecules on the polymer shell. The transition temperature of pure PNIPAM is 32°C, which is not useful for drug delivery applications due to it is placed under the physiological temperature. Fortunately, this temperature can be shifted up toward values located above the physiological conditions through the introduction of hydrophilic monomers within the polymeric chain. Guisasola et al. have employed *N*-(hydroxymethyl)acrylamide (NHMA) in combination with NIPAM in order to create a polymeric shell around MSN@SPION which suffered the transition at temperatures of 42–43°C, which are located in the hyperthermia range [51]. The magnetic responsiveness was provided by the SPION which were embedded within the silica that produced enough heat to trigger the polymer transition when the system was exposed to 838 kHz and $20.05\,kA\,m^{-1}$ alternating magnetic field. This nanodevice was able to release the housed drug even when the external temperature was maintained at physiological conditions (37°C) through its introduction into a thermostatic chamber. The AMF application triggered the drug departure without increasing the global temperature. The reason of this phenomenon is originated by the high local temperature increase which happens in the close proximity of the SPION because these magnetic cores act as *hot spots* which produces enough heat for causing the polymer collapse. Zink et al. have measured the temperature in the nanometric scale employing MSN which contained SPION and a photoluminescence NaYF4:Yb^{3+},Er^{3+} nanocrystal *co*-encapsulated in close proximity [52]. The intensity of the luminescence band registered was highly dependent of the temperature and thus, the application of AMF increased the temperature in the nanoscale altering the intensity of the emission band. The authors have reported that the temperature in the SPION surroundings can be increased up to 20°C more under AMF exposition at short times. This fact is tremendously important for the clinical application of this type of devices because, one important limitation of the systems based on magnetic hyperthermia is the necessity to employ large amounts of magnetic cores in order to increase the temperature in the target tissue. The demonstration of the fact that the *hot spot* effect can be employed for triggering the drug departure in MSN@SPION nanoparticles corresponds a strong support for the use of this type of nanodevices in the clinical field. Guisasola et al. have reported the application of MSN@SPION nanoparticles loaded with Dox and coated with the NIPAM/NHMA polymer as magnetically triggered drug release nanodevice in a murine model [53]. First, the authors studied the capacity to diffuse throughout the tumoral mass by the intratumoral injection of these nanoparticles labeled with rhodamine-B in order to monitorize their localization inside the tumor by fluorescence microscopy (Fig. 6A). The particles were observed in the surrounding of

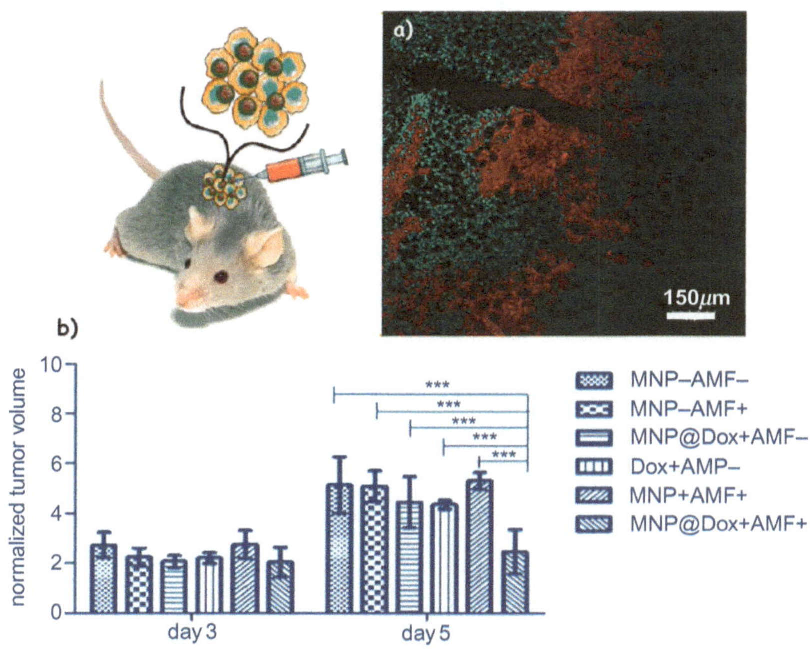

Fig. 6 (A) Fluorescence microscopy of rhodamine-labeled MMSN in the tumor after injection. (B) Tumor growth calculated as (tumor volume)/(initial tumor volume). Day 3 corresponds to the measurements after the last AMF treatment. Day 5 corresponds to 48 h after the last AMF treatment. $N=4$. $^{***}P<0.001$; two-way ANOVA, Bonferroni multiple comparison test.

the injection point but also in distant zones, showing good penetrability into the diseased tissue. Later, the capacity to induces therapeutic response with Dox-loaded particles (MMSN-DOXO) was evaluated by the injection and subsequent AMF exposition. In this study, the animals were divided in six groups: (1) control, (2) MMSN-DOXO without exposure to the AMF, (3) free Dox in the same concentration that the one loaded inside the silica, (4) AMF without nanoparticles, (5) MMSN-DOXO and AMF and (6) MMSN without Dox and exposed to AMF. The result of this experiment showed that only the group of mice treated with MMSN-DOXO and AMF exhibited strong tumor shrinkage whereas the rest of the groups did not present any effect (Fig. 6B).

5. CONCLUSIONS

As it has been widely described above, the use of responsive MSN which can be triggered by external stimuli such as US, light and magnetic fields represent a valuable tool for the controlled administration of potent therapeutic drugs. The nanometric size of these carriers allows the close interaction with the target tissues and cells achieving a significant enhancement in the selectivity of the treatment. Additionally, thanks to their stimuli-responsive properties, the administration of the housed drugs can be remotely triggered by the clinicians whose can control the dose with extremely high precision. The combination of these two properties into one single nanodevice, selectivity and control in the release, provides a powerful weapon in the fight against complex diseases as cancer which could reach the market in the coming years.

ACKNOWLEDGMENTS

This work was supported by European Research Council, ERC-2015-AdG (VERDI), Proposal No. 694160 and Ministerio de Economía y Competitividad (MINECO) (MAT2015-64831-R grant).

REFERENCES

[1] M. Vallet-Regi, A. Rámila, R.P. Del Real, J. Pérez-Pariente, A new property of MCM-41: drug delivery system, Chem. Mater. 13 (2001) 308–311, https://doi.org/10.1021/cm0011559.

[2] D. Tarn, C.E. Ashley, M. Xue, E.C. Carnes, J.I. Zink, C.J. Brinker, Mesoporous silica nanoparticle nanocarriers: biofunctionality and biocompatibility, Acc. Chem. Res. 46 (2013) 792–801, https://doi.org/10.1021/ar3000986.

[3] S. Mura, J. Nicolas, P. Couvreur, Stimuli-responsive nanocarriers for drug delivery, Nat. Mater. 12 (2013) 991–1003. https://doi.org/10.1038/nmat3776.

[4] A.K.W. Wood, C.M. Sehgal, A review of low-intensity ultrasound for cancer therapy, Ultrasound Med. Biol. 41 (2015) 905–928, https://doi.org/10.1016/j.ultrasmedbio.2014.11.019.
[5] S. Mitragotri, Healing sound: the use of ultrasound in drug delivery and other therapeutic applications, Nat. Rev. Drug Discov. 4 (2005) 255. https://doi.org/10.1038/nrd1662.
[6] J.E. Kennedy, High-intensity focused ultrasound in the treatment of solid tumours, Nat. Rev. Cancer 5 (2005) 321. https://doi.org/10.1038/nrc1591.
[7] A. Burgess, S. Dubey, S. Yeung, O. Hough, N. Eterman, I. Aubert, K. Hynynen, Alzheimer disease in a mouse model: MR imaging-guided focused ultrasound targeted to the hippocampus opens the blood–brain barrier and improves pathologic abnormalities and behavior, Radiology 273 (2014) 736–745, https://doi.org/10.1148/radiol.14140245.
[8] R.J.E. van den Bijgaart, D.C. Eikelenboom, M. Hoogenboom, J.J. Fütterer, M.H. den Brok, G.J. Adema, Thermal and mechanical high-intensity focused ultrasound: perspectives on tumor ablation, immune effects and combination strategies, Cancer Immunol. Immunother. 66 (2017) 247–258, https://doi.org/10.1007/s00262-016-1891-9.
[9] K. Hynynen, N. McDannold, N.A. Sheikov, F.A. Jolesz, N. Vykhodtseva, Local and reversible blood–brain barrier disruption by noninvasive focused ultrasound at frequencies suitable for trans-skull sonications, Neuroimage 24 (2005) 12–20. https://doi.org/10.1016/j.neuroimage.2004.06.046.
[10] D.O. Draper, J.C. Castel, D. Castel, Rate of temperature increase in human muscle during 1 MHz and 3 MHz continuous ultrasound, J. Orthop. Sport. Phys. Ther 22 (1995) 142–150, https://doi.org/10.2519/jospt.1995.22.4.142.
[11] C.D. Arvanitis, M. Bazan-Peregrino, B. Rifai, L.W. Seymour, C.C. Coussios, Cavitation-enhanced extravasation for drug delivery, Ultrasound Med. Biol. 37 (2011) 1838–1852. https://doi.org/10.1016/j.ultrasmedbio.2011.08.004.
[12] P. Cintas, S. Tagliapietra, M. Caporaso, S. Tabasso, G. Cravotto, Enabling technologies built on a sonochemical platform: challenges and opportunities, Ultrason. Sonochem. 25 (2015) 8–16. https://doi.org/10.1016/j.ultsonch.2014.12.004.
[13] S.R. Sirsi, M.A. Borden, State-of-the-art materials for ultrasound-triggered drug delivery, Adv. Drug Deliv. Rev. 72 (2014) 3–14. https://doi.org/10.1016/j.addr.2013.12.010.
[14] I.A.S. Elhelf, H. Albahar, U. Shah, A. Oto, E. Cressman, M. Almekkawy, High intensity focused ultrasound: the fundamentals, clinical applications and research trends, Diagn. Interv. Imaging 99 (2018) 349–359, https://doi.org/10.1016/j.diii.2018.03.001.
[15] Z. Yang, H. Xiaoxia, J. Xiangxiang, C. Yu, Construction of silica-based micro/nanoplatforms for ultrasound theranostic biomedicine, Adv. Healthcare Mater. 6 (2017) 1700646, https://doi.org/10.1002/adhm.201700646.
[16] H.-J. Kim, H. Matsuda, H. Zhou, I. Honma, Ultrasound-triggered smart drug release from a poly(dimethylsiloxane)–mesoporous silica composite, Adv. Mater. 18 (2006) 3083–3088, https://doi.org/10.1002/adma.200600387.
[17] A. Milgroom, M. Intrator, K. Madhavan, L. Mazzaro, R. Shandas, B. Liu, D. Park, Mesoporous silica nanoparticles as a breast-cancer targeting ultrasound contrast agent, Colloids Surf. B Biointerfaces 116 (2014) 652–657. https://doi.org/10.1016/j.colsurfb.2013.10.038.
[18] M. Ming, X. Huixiong, C. Hangrong, J. Xiaoqing, Z. Kun, W. Qi, Z. Shuguang, W. Rong, Y. Minghua, C. Xiaojun, L. Faqi, S. Jianlin, A drug–perfluorocarbon nanoemulsion with an ultrathin silica coating for the synergistic effect of chemotherapy and ablation by high-intensity focused ultrasound, Adv. Mater. 26 (2014) 7378–7385, https://doi.org/10.1002/adma.201402969.

[19] S.-F. Lee, X.-M. Zhu, Y.-X.J. Wang, S.-H. Xuan, Q. You, W.-H. Chan, C.-H. Wong, F. Wang, J.C. Yu, C.H.K. Cheng, K.C.-F. Leung, Ultrasound, pH, and magnetically responsive crown-ether-coated core/shell nanoparticles as drug encapsulation and release systems, ACS Appl. Mater. Interfaces 5 (2013) 1566–1574, https://doi.org/10.1021/am4004705.
[20] J.L. Paris, M.V. Cabanas, M. Manzano, M. Vallet-Regí, Polymer-grafted mesoporous silica nanoparticles as ultrasound-responsive drug carriers, ACS Nano 9 (2015) 11023–11033, https://doi.org/10.1021/acsnano.5b04378.
[21] H. Maeda, H. Nakamura, J. Fang, The EPR effect for macromolecular drug delivery to solid tumors: improvement of tumor uptake, lowering of systemic toxicity, and distinct tumor imaging in vivo, Adv. Drug Deliv. Rev. 65 (2013) 71–79, https://doi.org/10.1016/j.addr.2012.10.002.
[22] J. Shi, P.W. Kantoff, R. Wooster, O.C. Farokhzad, Cancer nanomedicine: progress, challenges and opportunities, Nat. Rev. Cancer 17 (2016) 20–37, https://doi.org/10.1038/nrc.2016.108.
[23] V.J. Venditto, F.C. Szoka, Cancer nanomedicines: so many papers and so few drugs! Adv. Drug Deliv. Rev. 65 (2013) 80–88, https://doi.org/10.1016/j.addr.2012.09.038.
[24] S. Wilhelm, A.J. Tavares, Q. Dai, S. Ohta, J. Audet, H.F. Dvorak, W.C.W. Chan, Analysis of nanoparticle delivery to tumours, Nat. Rev. Mater. 1 (2016) 16014, https://doi.org/10.1038/natrevmats.2016.14.
[25] Z. Gao, L. Zhang, J. Hu, Y. Sun, Mesenchymal stem cells: a potential targeted-delivery vehicle for anti-cancer drug loaded nanoparticles, nanomedicine nanotechnology, Biol. Med. 9 (2013) 174–184. https://doi.org/10.1016/j.nano.2012.06.003.
[26] J.M. Karp, G.S. Leng Teo, Mesenchymal stem cell homing: the devil is in the details, Cell Stem Cell 4 (2009) 206–216. https://doi.org/10.1016/j.stem.2009.02.001.
[27] Y.-L. Hu, Y.-H. Fu, Y. Tabata, J.-Q. Gao, Mesenchymal stem cells: a promising targeted-delivery vehicle in cancer gene therapy, J. Control. Release 147 (2010) 154–162. https://doi.org/10.1016/j.jconrel.2010.05.015.
[28] M. Roger, A. Clavreul, M.-C. Venier-Julienne, C. Passirani, L. Sindji, P. Schiller, C. Montero-Menei, P. Menei, Mesenchymal stem cells as cellular vehicles for delivery of nanoparticles to brain tumors, Biomaterials 31 (2010) 8393–8401. https://doi.org/10.1016/j.biomaterials.2010.07.048.
[29] M.I. Macias, J. Grande, A. Moreno, I. Domínguez, R. Bornstein, A.I. Flores, Isolation and characterization of true mesenchymal stem cells derived from human term decidua capable of multilineage differentiation into all 3 embryonic layers, Am. J. Obstet. Gynecol. 203 (2010) 495.e9–495.e23, https://doi.org/10.1016/j.ajog.2010.06.045.
[30] I. Vegh, M. Grau, M. Gracia, J. Grande, P. de la Torre, A.I. Flores, Decidua mesenchymal stem cells migrated toward mammary tumors in vitro and in vivo affecting tumor growth and tumor development, Cancer Gene Ther. 20 (2012) 8. https://doi.org/10.1038/cgt.2012.71.
[31] J.L. Paris, P. De La Torre, M. Victoria Cabañas, M. Manzano, M. Grau, A.I. Flores, M. Vallet-Regí, Vectorization of ultrasound-responsive nanoparticles in placental mesenchymal stem cells for cancer therapy, Nanoscale 9 (2017) 5528–5537, https://doi.org/10.1039/c7nr01070b.
[32] J.L. Paris, P.D.L. Torre, M. Manzano, M.V. Cabañas, A.I. Flores, M. Vallet-Regí, Decidua-derived mesenchymal stem cells as carriers of mesoporous silica nanoparticles. In vitro and in vivo evaluation on mammary tumors, Acta Biomater. 33 (2016) 275–282, https://doi.org/10.1016/j.actbio.2016.01.017.
[33] Y. Chen, Q. Meng, M. Wu, S. Wang, P. Xu, H. Chen, Y. Li, L. Zhang, L. Wang, J. Shi, Hollow mesoporous organosilica nanoparticles: a generic intelligent framework-hybridization approach for biomedicine, J. Am. Chem. Soc. 136 (2014) 16326–16334, https://doi.org/10.1021/ja508721y.

[34] X. Qian, W. Wang, W. Kong, Y. Chen, Organic-inorganic hybrid hollow mesoporous organosilica nanoparticles for efficient ultrasound-based imaging and controlled drug release, J. Nanomater. 2014 (2014).
[35] N.K. Mal, M. Fujiwara, Y. Tanaka, Photocontrolled reversible release of guest molecules from coumarin-modified mesoporous silica, Nature 421 (2003) 350–353, https://doi.org/10.1038/nature01337.1.
[36] Y. Zhu, M. Fujiwara, Installing dynamic molecular photomechanics in mesopores: a multifunctional controlled-release nanosystem, Angew. Chem. Int. Ed. 46 (2007) 2241–2244, https://doi.org/10.1002/anie.200604850.
[37] D. He, X. He, K. Wang, J. Cao, Y. Zhao, A light-responsive reversible molecule-gated system using thymine-modified mesoporous silica nanoparticles, Langmuir 28 (2012) 4003–4008, https://doi.org/10.1021/la2047504.
[38] N.Z. Knezevic, B.G. Trewyn, V.S.Y. Lin, Functionalized mesoporous silica nanoparticle-based visible light responsive controlled release delivery system, Chem. Commun. 47 (2011) 2817–2819, https://doi.org/10.1039/c0cc04424e.
[39] J. Lai, X. Mu, Y. Xu, X. Wu, C. Wu, C. Li, J. Chen, Y. Zhao, Light-responsive nanogated ensemble based on polymer grafted mesoporous silica hybrid nanoparticles, Chem. Commun. 46 (2010) 7370–7372, https://doi.org/10.1039/C0CC02914A.
[40] M. Martinez-Carmona, A. Baeza, M.A. Rodriguez-Milla, J. García-Castro, M. Vallet-Regí, Mesoporous silica nanoparticles grafted with a light-responsive protein shell for highly cytotoxic antitumoral therapy, J. Mater. Chem. B 3 (2015) 5746–5752, https://doi.org/10.1039/C5TB00304K.
[41] S. Yang, N. Li, D. Chen, X. Qi, Y. Xu, Y. Xu, Q. Xu, H. Li, J. Lu, Visible-light degradable polymer coated hollow mesoporous silica nanoparticles for controlled drug release and cell imaging, J. Mater. Chem. B 1 (2013) 4628, https://doi.org/10.1039/c3tb20922a.
[42] M. Martínez-Carmona, D. Lozano, A. Baeza, M. Colilla, M. Vallet-Regí, A novel visible light responsive nanosystem for cancer treatment, Nanoscale 9 (2017) 15967–15973, https://doi.org/10.1039/C7NR05050J.
[43] G. Sahay, D.Y. Alakhova, A.V. Kabanov, Endocytosis of nanomedicines, J. Control. Release 145 (2010) 182–195, https://doi.org/10.1016/j.jconrel.2010.01.036.
[44] J. Liu, C. Wang, X. Wang, X. Wang, L. Cheng, Y. Li, Z. Liu, Mesoporous silica coated single-walled carbon nanotubes as a multifunctional light-responsive platform for cancer combination therapy, Adv. Funct. Mater. 25 (2015) 384–392, https://doi.org/10.1002/adfm.201403079.
[45] E. Guisasola, A. Baeza, L. Asín, J.M. Fuente, Heating at the nanoscale through drug-delivery devices: fabrication and synergic effects in cancer treatment with nanoparticles, Small Methods (2018), https://doi.org/10.1002/smtd.201800007.
[46] L. Hui, L.-L. Tan, P. Jia, Q.-L. Li, Y.-L. Sun, J. Zhang, Y.-Q. Ning, J. Yu, Y.-W. Yang, Near-infrared light-responsive supramolecular nanovalve based on mesoporous silica-coated gold nanorods, Chem. Sci. 5 (2014) 2804–2808, https://doi.org/10.1039/c4sc00198b.
[47] G. Villaverde, S. Gómez-graña, E. Guisasola, I. García, C. Hanske, L.M. Liz-marzán, A. Baeza, M. Vallet-regí, Targeted chemo-photothermal therapy: a nanomedicine approximation to selective melanoma treatment, Part. Part. Syst. Charact. (2018), https://doi.org/10.1002/ppsc.201800148.
[48] P. Saint-Cricq, S. Deshayes, J.I. Zink, A.M. Kasko, Magnetic field activated drug delivery using thermodegradable azo-functionalised PEG-coated core–shell mesoporous silica nanoparticles, Nanoscale 7 (2015) 13168–13172, https://doi.org/10.1039/C5NR03777H.
[49] E. Ruiz-Hernández, A. Baeza, M. Vallet-Regí, Smart drug delivery through DNA/magnetic nanoparticle gates, ACS Nano 5 (2011) 1259–1266, https://doi.org/10.1021/nn1029229.

[50] A. Baeza, E. Guisasola, E. Ruiz-Hernández, M. Vallet-Regí, Magnetically triggered multidrug release by hybrid mesoporous silica nanoparticles, Chem. Mater. 24 (2012) 517–524, https://doi.org/10.1021/cm203000u.
[51] E. Guisasola, A. Baeza, M. Talelli, D. Arcos, M. Moros, J.M. de la Fuente, M. Vallet-Regí, Magnetic-responsive release controlled by hot spot effect, Langmuir 31 (2015) 12777–12782, https://doi.org/10.1021/acs.langmuir.5b03470.
[52] J. Dong, J.I. Zink, Taking the temperature of the interiors of magnetically heated nanoparticles, ACS Nano 8 (2014) 5199–5207, https://doi.org/10.1021/nn501250e.
[53] E. Guisasola, L. Asín, L. Beola, J.M. de la Fuente, A. Baeza, M. Vallet-Regí, Beyond traditional hyperthermia: in vivo cancer treatment with magnetic-responsive mesoporous silica nanocarriers, ACS Appl. Mater. Interfaces 10 (2018) 12518–12525, https://doi.org/10.1021/acsami.8b02398.

CHAPTER SIX

The Bioimaging Applications of Mesoporous Silica Nanoparticles

Feby Wijaya Pratiwi*,†,‡, Chiung Wen Kuo†, Si-Han Wu§, Yi-Ping Chen§, Chung Yuan Mou‡, Peilin Chen†,1

*Nanoscience and Technology Program, Taiwan International Graduate Program, Taipei, Taiwan
†Research Center for Applied Sciences, Academia Sinica, Taipei, Taiwan
‡Department of Chemistry, National Taiwan University, Taipei, Taiwan
§Graduate Institute of Nanomedicine and Medical Engineering, Taipei Medical University, Taipei, Taiwan
[1]Corresponding author: e-mail address: peilin@gate.sinica.edu.tw

Contents

Abstract

The unique features of Mesoporous Silica Nanoparticles (MSNs) provide a suitable platform to carry fluorescence dyes for various bioimaging applications. Several strategies have been developed to conjugate a variety of dyes either in the pores or on the surfaces of MSNs to form the fluorescence MSNs (FMSNs). In this chapter, we will discuss recent research progress and future development of FMSNs for living system imaging. We will first describe different strategies for the fabrications of FMSNs. Then, we will discuss the recent developments of cellular and intracellular imaging including self-probe for the interactions of FMSNs with the cells, receptor and organelle labeling, sensing and tracking of biological system, and monitoring the drug delivery and release processes. Moreover, we will include the applications of FMSNs as contrast agents for in vivo imaging. Finally, we will conclude and highlight the challenges and opportunities for MSNs in medical applications.

The Enzymes, Volume 43
ISSN 1874-6047
https://doi.org/10.1016/bs.enz.2018.07.006

1. INTRODUCTION

Nanoparticles have been developed for many biomedical applications such as early diagnostics tools or therapeutic agent carriers. Ironically, only a few of them successfully enter into the market. Despite the "promising" efficacy issues, the fates of the nanocarriers and environmental influences of therapeutic compounds inside biologic systems are not fully understood due to their complexity, diversity and dynamic activities in the living system. To better understand the behavior of nanoparticles inside the living system, it is desirable to directly visualize the nanoparticles in living systems in their native states. However, in vivo visualization should be studied noninvasively to provide reliable results. Therefore, it requires quantitative tools to monitor the biological events and to provide statistical analysis to validate the hypothesis through various studies.

To investigate the biological processes in the living system, optical imaging offers several advantages including easy-of-use, simultaneous study of multi-subjects through labeling, high spatiotemporal resolution, and direct data interpretation through visualization. Recent advances in optical imaging allow us to visualize the subcellular structures with a spatial resolution breaking the diffraction limit. These optical tools include Stimulated Emission Depletion (STED) microscopy [1, 2], Structure Illumination Microscopy (SIM) [3], Photoactivated Localization Microscopy (PALM) [4, 5], and Stochastic Optical Reconstruction Microscopy (STORM) [6, 7]. On the other hand, the development of high power laser and fast detection system enables the real-time imaging of individual cells in the living animals through multiphoton microscopy.

One important issue in optical imaging is the photon budget where the probing system should provide enough photons for high-quality imaging. The searching and improving the design of imaging probes, which could provide the high number of photons at the targeted sites, are essential. Furthermore, for imaging living systems such as living cells and animals, it is very important to engineer the probes with desired functionality while reducing the long-term toxicity for the living targets. To achieve these criteria, the design of probes should meet the following requirements: (a) the size of probes should be small enough for cellular uptake, (b) the structure of probes should be stable both in culture medium and circulation, (c) the probes should selectively interact with the targeted organelles or cells while maintaining minimum disturbance to the living system, (d) the probes should provide enough photon output for reliable tracking and statistical analysis, and (e) the response of probes should be fast enough to faithfully report the rapid changes of biological processes.

In pursuing the abovementioned requirements, one strategy is to secure the fluorescence probes within the nanoparticle matrix such as Mesoporous Silica Nanoparticles (MSNs). The unique properties of MSNs such as tunable pore size and high pore volume/surface area provide high loading capacity of various fluorescence molecules while protecting the dye molecules from the harsh environmental factors resulting in enhanced photostability and concentrated optical signals with better spatiotemporal resolution [8, 9]. Moreover, the surfaces of fluorescent MSNs (FMSNs) can be engineered with the targeting ligands and various surface modifications to improve their stability, selective recognition, which potentially allow multi-labeling or co-delivery of therapeutic agents in the living systems.

In this chapter, we will review the recent research progress and future development of the bioimaging using MSNs. We will first outline the development of various types of FMSNs and the roles of surface functionalization in their interactions and targeting in the biological systems. We will then discuss recent progresses for cellular and intracellular imaging, which includes self-probe for the interactions of FMSNs with the cells, receptor and organelle labeling, sensing and tracking of biological system, and monitoring the drug delivery and release processes. We will also describe the promising applications of FMSNs as imaging probes for nanoparticles distribution, cell tracker, contrast agent for tumor imaging and theranostics. Finally, we will conclude and highlight the challenges and opportunities for MSNs in medical applications.

2. PREPARATION OF FLUORESCENCE DYES FUNCTIONALIZED MSNs

2.1 General Synthesis of Mesoporous Silica Nanoparticles

First successful synthesis of mesoporous silica nanoparticles (MSN was introduced by Victor Lin [10]), was reported by several groups [11–13]. The synthesis process was based on the sol–gel process consisting of the hydrolysis of tetraethyl orthosilicate (TEOS) and polymerization to form Si–O–Si bonds around a surfactant mesophase, which was then followed by the removal of surfactant template using calcination or chemical extraction, resulting in the mesoporous materials. Over the years, with a deeper understanding

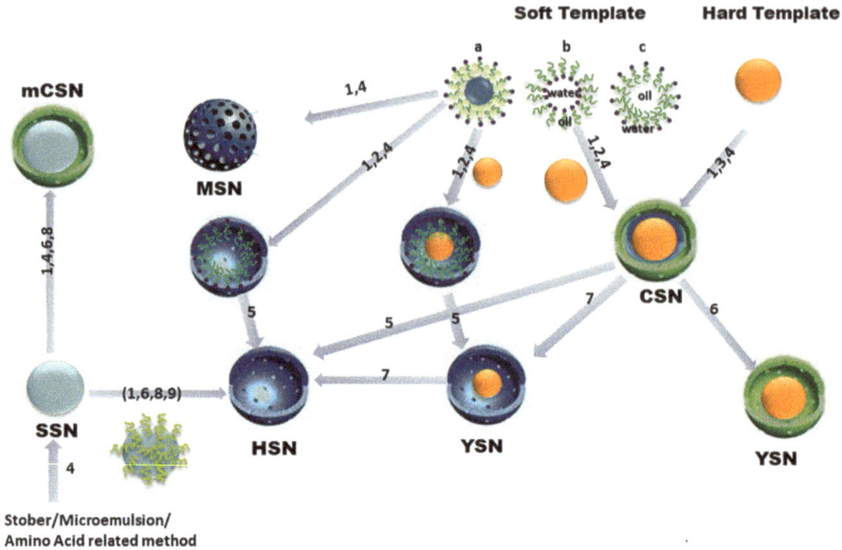

Fig. 1 Schematic illustration of the different synthesis of various types of MSN.

of sol–gel process and increase of availability of different types of surfactants, MSNs with various structures and morphologies, pore sizes and structures, dimensions, and multifunctional counterparts have been developed. The demands for MSNs with special structures and performance, which could be applied in the biomedical fields, are gradually increasing. For the bio-imaging applications, different types of MSNs: ordered mesoporous silica (OMS), hollow and rattle-type MSNs, and core/shell structure MSNs have been developed with different kinds of synthesis strategies. Different types of MSNs synthesis schemes are illustrated in Fig. 1.

1. Surfactant self-assembly
2. Micelle, emulsion, bubble gas, polymers
3. Stöber and microemulsion methods
4. Silica spheres nanoparticle template
5. Soft/hard template removal
6. Structural different based selective etching
7. Compositional different based selective etching
8. Surface protection
9. Dissolution, redeposition, and self-transformation

2.1.1 Ordered Mesoporous Silica (OMS)

MCM-41 is one type of MSNs which possess uniform arrangement of two-dimensional hexagonal pore structures [14]. Generally, for MCM-41 synthesis (Fig. 2), the cationic surfactants such as cetyltrimethylammonium

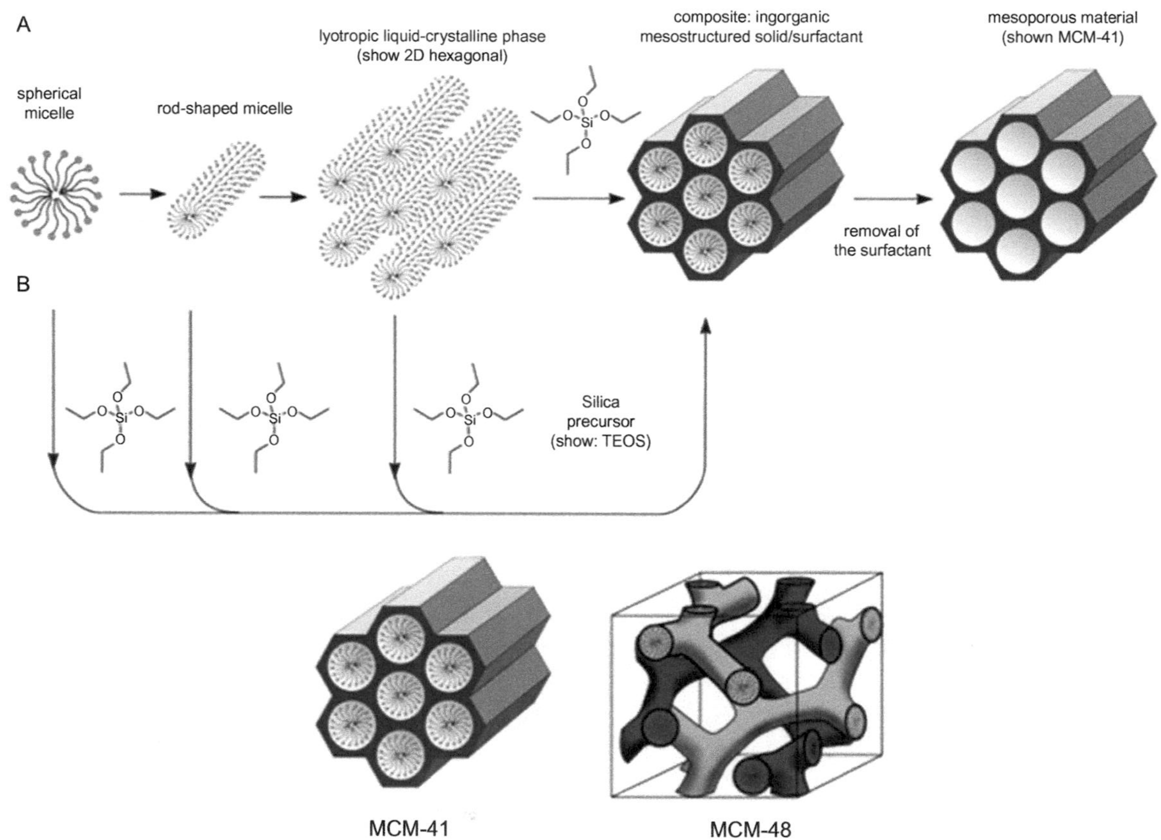

Fig. 2 Top: the schematic of MCM-41 preparation by (A) the first proposed model and (B) the second proposed model and bottom: different structure of mesoporous MCM-41 (2D hexagonal, space group p6mm) and MCM-48 (cubic, space group Ia3d) [15].

bromide (CTAB) are used for the soft templating. In this process, the CTAB molecules first form a micelles template as their concentration is above the critical micelle concentration, which is then followed by condensation of silica precursors TEOS or sodium metasilicate (Na_2SiO_3) around polar head region organic micelles under the alkaline environment [15]. After removal of templates, MCM-41 can be obtained with a tunable pore size within the range of 1.6–10 nm and a surface area higher than 700 m^2/g. Tuning the molar ratio of surfactants and silica sources and tailoring the pH of base catalyst or by cooperating the organoalkoxysilane precursors during the co-condensation reactions could produce MSNs with different size, morphology, and shape (sphere-, rod-, and wormlike structures). In order to obtain monodispersed MSNs with narrow distribution size, Mou's group separated the synthesis process into two steps: nuclei formation and particle growth. In this strategy, the pH value of the solution is an essential parameter to control the condensation rate which influences the number of nuclei and the size of MSNs. These MSNs were efficiently uptaken by cells and were used for a cell marker [16]. Hiroaki et al. [17] and Kaszpurenko et al. [18] reported physically entrapment of organic dyes inside the silica matrix of MCM-41 through a one-step self-assembly process. Hydrophobic dyes or the hydrophilic dyes could be incorporated to the hydrophobic cores of the surfactant micelles via electrostatic and/or hydrophobic interactions and encapsulated in the silica matrix.

MCM-48 is another type of MSNs, which have three-dimensional bicontinuous penetrating channel structures associated to the space group of cubic Ia3d symmetry. At the beginning of the development of MCM-48, the synthesis process used cationic–anionic surfactant template, which required high temperature and a long time reaction resulting in nanoparticle's size larger than 1 μm. In the past few years, Lin and co-worker successfully obtained monodispersed spherical MCM-48 based on the modified Stöber method at room temperature. In this approach, the size of spherical MCM-48 could be controlled (ranging from 70 to 500 nm) by tuning the concentrations of surfactant template, triblock copolymer Pluronic F127 [19]. For the applications, MCM-48 has been reported to carry the methylene viologen units on the pore walls of the mesopores via co-hydrolysis and co-condensation of dichloride of *N*,*N'*-bis(triethoxysilylmethyl)-4,4′-bipyridinium (VP) and TEOS [20].

2.1.2 Hollow/Rattle-Type Mesoporous Silica Nanoparticles

"Yolk–shell" or "rattle-type" nanostructures are an important and interesting group in the complex hollow nanoarchitectures because of their unique capability to provide a controllable space between the core and the outer shell of MSNs. The hollow interior spaces and shell pores offer high loading capacity of different agents allowing multi-functionalities or co-delivery of several imaging probes as well as therapeutic agents. Similar to the regular MSNs, the synthesis includes the sol–gel process, which allows silica matrix to grow to surround the template, and subsequent template removal. In the hollow type of MSNs synthesis, hard and soft templates are commonly used to generate interstitial hollow spaces and shell pores, respectively. On the other hand, the synthesis of rattle-type MSNs can be achieved by multilayer coating on the cores of the particles to create the removal middle layer.

The hard templating method is recently considered as a better synthesis strategy in order to achieve high-quality hollow/rattle MSNs in term of monodispersity and reproducibility. In these approaches, the selection of hard templates and etchants is crucial, which highly depends on the etching mechanisms. Various materials such as silica, iron oxide nanoparticles, polystyrene beads, etc. have been employed as hard templates, while agents such as sodium carbonate (Na_2CO_3), sodium hydroxide (NaOH), hydrofluoric acid (HF), hydrochloric acid (HCl), and even hot water have been utilized as etchants. Cho et al. have utilized latex beads or polystyrene (100 nm) as a core template combined with CTAB as the soft co-template to create mesoporous hollow silica nanoparticles (mHSNs). After the forming the mHSN, amine groups ($–NH_2$) were grafted on the surfaces just before the calcination to obtain mHSN-NH_2.The amine group on the surfaces provided the reaction site for the PEGylation, which was used to improve the dispersion of nanoparticles in aqueous solutions. These mHSNs were loaded with the indocyanine green to form an activatable and selective theranostic platform for cancer imaging using near-infrared (NIR) [21].

Mou and co-workers [22] developed an alternative approach for the fabrication of hollow and yolk/shell silica nanospheres by W/O (water/oil) microemulsion-template method. In this process, the microemulsion template was formed from the mixture of cyclohexane, Triton X-100, hexanol, and water. Moreover, the microemulsion could contain pre-made oil-soluble nanoparticles like Fe_2O_3 or Qdots. After forming the microemulsion template, the silica was polymerized around the template to produce a mesoporous layer. This process allowed the encapsulation of nanoparticles inside the final hollow silica spheres. The particle size and the interior space could be easily tuned by controlling the order and the amount of APTMS additions. This microemulsion method is preferred to produce MSNs with the size <100 nm. Dual or multi-surfactants have been used as a soft template to simultaneously form the interstitial hollow and porous shell. Xu's group have developed one-step reaction to fabricate Yolk/SiO_2 shell by adding the SiO_2 or Au or Fe_3O_2 nanoparticles, which acted as cores in the mixtures of a zwitterionic surfactant (lauryl sulfonate betaine (LSB)) and anionic surfactant (sodium dodecyl benzene sulfonate (SDBS)) [23]. For imaging applications, yolk–shell upconversion nanoparticles (YSUCNP) with $NaYF_4$:Yb/Er@ $NaGdF_4$ as moveable core and silica mesopores as shell layer were synthesized via a seeded growth approach (UCNPs@SiO_2). Using similar aqueous phase regrowth technique, the silica sources polymerized around the CTAB template resulting in mesoporous layer on UCNPs@SiO_2 core. After removing inner silica, the UCNPs–yolk–shell nanoparticles with spacious interior hollow were formed. The large pore volume of the hollow nanoparticles facilitated the enormous amount of DOX loading for chemotherapy whereas the Gd-based upconversion yolk served as dual-modal UCL/MR imaging probe. Such UCNPs–yolk–shell nanoparticles allowed dual-modal imaging for visualizing the targeting and effects of chemotherapy [24]. Similarly, lanthanide upconversion nanophosphor was employed as a movable core in rattle mesoporous silica shell. In the hollow cavities, the colorimetric dyes were used for cysteine/homocysteine intracellular sensing based on luminescent resonance energy transfer (LRET) in a UCL turn-off pattern [25].

An alternative self-template technique for the synthesis of hollow type MSNs was developed without extra "template." The treatment with the alkaline solution could self-transform the MSN spheres which protected by cationic polyelectrolyte (poly-(dimethyldiallylammonium chloride) (PDDA)) layer into the hollow structures [26]. Under ammonia treatment, the hydroxyl ions entered the PDDA to react with the silica sphere, which generated silicate oligomers. The negatively charge silicate oligomers tended to interact and deposit on the PDDA layer possessing positive charges, cross-linked and finally formed silica–PDDA complex shell. Yin and co-workers utilized poly(vinylpyrrolidone) (PVP) as the protecting layer. Under the $NaBH_4$ treatment, solid silica spheres self-transformed into the rattle-type structure and finally formed the hollow structure at a mild temperature [27].

2.1.3 Core/Shell Type Mesoporous Silica Nanoparticles

Fluorescence MSNs can also be obtained by growing the mesoporous shells on the fluorescent solid cores such as gold, Qdots, UCPN, or other organics fluorescence dyes to form core/shell particles to improve the fluorescence properties. Generally, two methods have been used to fabricate the core/shell types of MSNs namely Stöber sol–gel chemistry and microemulsion. For instance, Zhang and co-worker demonstrated that multifluorescent core/shell MSNs can be obtained in a three-step process. First, three different dyes, fluorescein isothiocyanate (FITC), rhodamine B isothiocyanate (RBITC), and rhodamine 101 succinimide (R101-SE), were conjugated with 3-aminopropyltriethoxysilane (APTS). Second, the hydrolysis and co-condensation of dye-conjugated APTS and silica source were performed to produce multicolor dye-doped silica cores (DSC). Finally, via the modified Stöber method, the mesoporous layer was deposited on the surfaces of DSCs to fabricate the shell. These core/shell nanoparticles exhibited very bright multicolor emission and could produce effective FRET under single wavelength excitation. Moreover, as these core/shell MSNs possessed numerous pores in the shell, the nanoparticles could effectively carry drug and simultaneously served as contrast agents for in vivo imaging to monitor multiple-target drug transport [28]. In 2006, Hyeon group reported core/shell nanoparticles could be obtained by coating the hydrophobic inorganic nanoparticles with MSNs. In these methods, the hydrophobic CdSe/ZnS QDs were first transferred into aqueous phase via forming microemulsions in the presence of CTAB surfactants. Then, the mesoporous silica was formed around the CTAB/QD complex cores via a silica sol–gel reaction. Using this method, the thickness of the silica shell could easily be tailored [29]. The Hyeon method was considered easier compared to the traditional Stöber sol–gel chemistry and microemulsion methods, which were widely used in many subsequent studies.

2.2 Incorporated the Fluorescence Dyes With MSNs

The silica chemistry allows easy modifications of MSNs with various functional groups (e.g., phosphonate (CH_3HPO_2), carboxyl (COOH), hydroxyl (OH), amine (NH_2), poly(ethylene glycol) (PEG)) via silane chemistry to facilitate further conjugation. There are many approaches for preparing the fluorescence mesoporous silica nanoparticles (FMSNs), including grafting organic fluorescent dyes onto the surfaces of the mesoporous framework of MSNs, encapsulating dyes directly within the silica matrix by co-condensation methods, fabricating core–shell structures, and physical adsorption methods. In this section, we will discuss some examples of the integration of various types of fluorophores into MSNs.

The organic dyes–MSNs are often used for imaging applications. Common strategies for preparing hybrid organic dyes–MSNs include incorporating

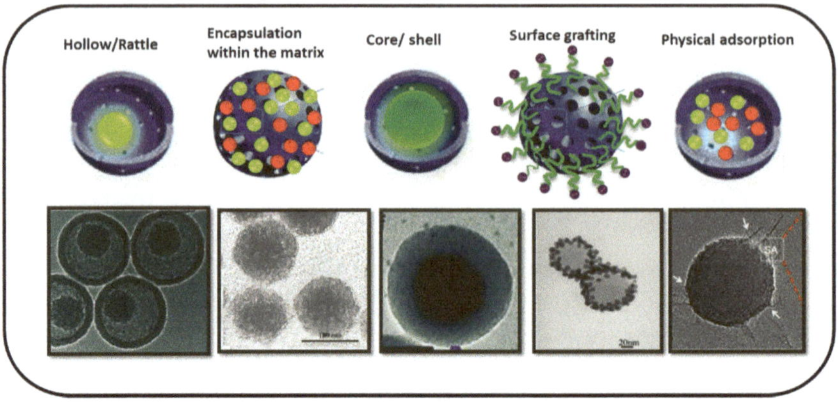

Fig. 3 General strategy for incorporating the fluorescence dyes with MSN.

organics dyes into MSNs either by grafting on the surfaces of MSNs via chemoselective chemistry [30, 31] or encapsulating them into silica matrix (Fig. 3) [32, 33]. For example, the silanol groups on the mesoporous surfaces could mediate the covalent conjugation of organics dyes via forming amine-modified MSNs. To form the FMSNs, FITC were grafted onto amine-modified MSNs. The FMSNs were then decorated with the polyvinyl pyridine (PVP) to maintain the colloidal stability [31]. Such covalent linkage is considered to be simple and stable chemical approaches to produce FMSNs. In another encapsulation strategy, triethoxysilylated fluorescein derivatives were firstly prepared by conjugating the FITC with 3-aminopropyl triethoxysilane. The TEOS and triethoxysilylated fluorescein in the W/O microemulsion mediated alkoxide hydrolysis were then condensed to form the FMSNs. The surfaces of FMSNs were grafted with different targeting ligands, which could specifically target the overexpressed receptors on the tumor cells. This approach has been used to monitor the interaction and the targeting capability of FMSNs in tumor cells by optical imaging [30]. However, the applications of a visible range of organic fluorescence dyes in the tissue level are limited due to poor light penetration depth and the interference of autofluorescence. To overcome these problems, Lee et al. encapsulated the indocyanine green (ICG) into MSNs via electrostatic interactions to fabricate the MSN-based NIR probes [34]. These strategies successfully enhanced the fluorescence photon count of ICG. Similarly, the squaraine (zwitterionic IR dyes) were also loaded onto the MSN's pores via physical adsorption and then followed by wrapping the pores entrance with graphene oxide (GO) to prevent the dye from unwanted nucleophilic substitution and leakage [35].

Inorganic semiconductor quantum dots (QDs) offer several advantages over the organic fluorescence. For example, the emission of QDs can be tailored within the range visible to infrared by controlling the physiochemical properties of QDs such as size and composition. Other properties such as broader excitation spectra with high absorption coefficients, high photostability, high quantum yield, and low photobleaching make QDs excellent optical contrast agents. However, the concerns of potential toxicity due to the leaching of heavy metal ions from QDs prevent their applications in biomedical imaging. Growing a thin layer of silica mesoporous on the QDs may provide a solution to overcome this concern. Therefore, many strategies have been developed to protect QDs within a mesoporous silica shell [10, 29, 36]. For example, quantum dots coated MSNs were synthesized via a sol–gel process in the presence of CTAB surfactants to form a microemulsion [29]. To improve QDs stability in the water, Pan et al. incorporated their core–shell (CdSe/ZnS)/MSNs with PEGylated liposome layer. After the coating, a slight redshift from 621 to 646 nm was observed. Sathe et al. sequentially and simultaneously loaded the hydrophobic CdSe/ZnS and magnetic iron oxide nanoparticles into MSNs via hydrophobic interactions of the octadecyl (C18) chain to fabricate a dual function probe. The amphiphilic polymer octylamine poly(acrylic acid) was grafted on the surfaces to enhance their biocompatibility. These MSNs could be used for the capture and enrichment of biological targets by applying the magnetic field and simultaneously acted as ratiometric sensing probes [36].

Carbon dots (CDs) is another material, which holds great promising for imaging probes due to their favorable characteristics such as high photostability, resistance to the photobleaching, less toxic, and broad two-photon excitation spectra. Recently, several approaches have been reported for incorporating the CDs with MSNs. For instance, Lei et al. fabricated ultra-small silylated CDs through pyrolysis of citric acid in the presence of coupling agent with *N*-(β-aminoethyl)-γ-aminopropyl methyldimethoxy silane (AEAPMS) at 240 °C. The CD-based-FMSNs were synthesized by the co-condensation of silylated CDs and TEOS in the NaOH environment. The presence of CTAB surfactants acted as templates and assisted the CDs to reside into mesopore walls. This CD-MSN emitted strong green fluorescent, which potentially can be utilized for a ratiometric sensor, cell imaging, and multichannel traceable drug delivery vehicles [37]. A unique fabrication of MSNs capped CDs was introduced by Pandey et al. using the pyrolyzed rice husks as silica sources. In the mixture of CTAB, NaOH, and ethanol, the pyrolyzed rice husks were spontaneously polymerized forming the MSNs without any additional silica precursor. The MSNs capped CDs exhibited strong fluorescence for labeling yeast cells [38]. Furthermore, the impregnated MSNs, citric acid, and ethylenediamine were treated with hydrothermal approaches for embedding and in situ growth of CDs inside the MSN pores (CDs@MSNs). The CDs@MSNs emitted stronger blue fluorescence compared to free-form of CDs, which were intravenously administrated with (for 1, 4 and 12h) in the tissues of normal mice under excitation of 350nm.

Rare-earth upconversion nanoparticles (UCNPs) pose several advantages compared to down-conversion fluorescent materials such as the capability of emitting higher-energy photons after absorbing two or more lower-energy photons with low background, penetrating into deep tissues and low phototoxicity to biological tissues. For instance, a layer of mesoporous silica was grown on the core of NaYF4:Tm/Yb/Gd UCNPs (NaYF4:Tm/Yb/Gd@MSNs) by phase transfer silica sol–gel reaction resulting in size particle smaller than 50nm. The NaYF4:Tm/Yb/Gd cores emitted IR emission at 800nm (ex: 980nm) while simultaneously the Gd^{3+} ions acted as a T1 MRI contrast agent. The NaYF4:Tm/Yb/Gd@MSNs were successfully tested in MCF-7 human breast cancer as well as in the mouse. By using NIR excitation, the emission of the nanoparticles was observed in the cytoplasm. When NaYF4:Tm/Yb/Gd@MSNs were intravenously injected into the mouse, the luminescence signal in the body could be observed [39].

3. FMSNs FOR IMAGING AT CELLULAR LEVELS

3.1 A Self-probe to Investigate the Cellular Interactions of MSNs

Understanding the interactions of functionalized MSNs and cells are important issues to improve the design rational for the cellular and intracellular applications. The size, shape and functional groups of MSNs have been demonstrated to be "standard" parameters influencing cellular interactions and uptake. Moreover, the uptake of nanoparticles has been shown to be concentration-dependent, where sigmoid function behavior of dose–response endocytosis [40], time-dependent, and cell-type dependent processes were observed [41].

Mou's group has observed the size-dependent cellular uptake of spherical FITC-MSNs (30, 50, 110, 170, and 280nm). It was shown that the intracellular uptake was highly particle-size-dependent with optimum size at 50nm (50 > 30 > 110 > 280 > 170nm) [42]. This optimal size at around 50nm for intracellular uptake was also observed in other nanoparticles [43, 44]. Indeed, the size of viruses and lipid-carrying proteins are within this range indicating that these sizes are probably the most favorable for cell endocytosis. The sizes of MSNs not only influence their uptake capacity but also affect the membrane trafficking of endosomal vesicles. Aoyama et al. quantitatively examined the intracellular trajectories and mobilities of FMSNs with different diameters inside the cells using single particle tracking technique [45]. Two major results were reported by this group. First, the mobility of nanoparticles was faster than particles with sub-micrometer sizes. More importantly, it was found that the nanoparticles did not suppress the speed of endosomal vesicle transport, whereas sub-micrometer particles perturbed the endosomal vesicle transport.

In addition to the size and the shape of MSNs, it is known that both surface potential and modifications on the surfaces of MSNs can affect the endocytic trafficking in cells. Incorporating with two fluorophores, FITC and rhodamine B isothiocyanate (RITC) for ratiometric pH sensing, the effect positively charged MSNs (functionalized with *N*-trimethoxysilylpropyl-*N*,*N*, *N*-trimethylammonium chloride, MSN-TA) and negatively charged MSNs (functionalized with 3-trihydroxysilylpropylmethylphosphonate, MSN-PP) have been utilized to investigate the MSNs uptake process for HeLa cells. Considering the membrane properties, the results suggested that nanoparticles were able to pass the membrane via a charge mediated membrane–nanoparticle interaction mechanism and internalized via a clathrin- and actin-mediated endocytosis. Moreover, positively charged MSN-TA mostly resided in higher pH regions, mainly in the cytosol whereas the negatively charged MSN-PPs were likely trapped in acidic endosomes [46]. Similarly, Lin and co-workers investigated different uptake mechanism of FITC-MSNs (MCM-41), which were grafted with four different ligands in HeLa cells: 3-aminopropyl (AP), *N*-folate-3-aminopropyl (FAP), guanidinopropyl (GP), and 3-[*N*-(2-guanidinoethyl)guanidino] propyl (GEGP) [40]. It was demonstrated that the uptake of unmodified and FAP modified FITC-MSNs was via a clathrin-mediated pathway, which depended on av-ATPase-dependent transport mechanism and also involved a dynamic microtubule network. In addition, the FAP was also internalized via folic acid receptors, which enhanced the efficiency of the endocytic process [47]. On the other hand, AP- and GP-MSNs were endocytosed via a caveolae-mediated mechanism [40]. Furthermore, the conjugation of MSNs with Tat-derived peptides with noncanonical Arg shortened side chain length was reported to enhance the energy-independent non-endocytotic pathways such as direct transduction [48]. More efforts have been taken to enhance the cytosolic delivery and to avoid the endosomal trapping by coating the MSNs, which carried Atto488 labeled cyt C, with lipids. The results suggested that the internalization of MSNs shifted from endocytosis to a direct cytosolic entry via membrane fusion [49].

Moreover, the interactions and intracellular behavior of MSNs are also depending on the types of the cells. The MSNs have been shown to be a promising probe for stem cell labeling/tracing. The unique character of the stem cell is the self-renewal through cell division and the potency (the capacity to differentiate into different specialized cell types). With the development of innovative personalized-regenerative technology using these stem cells, one essential aspect is the biocompatibility of functionalized FMSNs with stem cells. Mou and co-worker have investigated the effect of different surface modification of MSNs (using a different degree of *N*-trimethoxysilylpropyl-*N*,*N*,*N*-trimethylammonium chloride on their surface) to the proliferation, differentiation, and viability of human mesenchymal stem cells (hMSCs) and 3T3-L1 cells. The results indicated that the functionalization of MSNs could enhance their uptake into the stem cells. While the uptake of unmodified-MSNs in all stem cells was regulated by a clathrin-mediated pathway, the strongly positive charge MSNs only exhibited the same mechanism in 3T3-L1 cells. Regardless of the surface modifications, MSNs internalization did not affect cell viability, growth, adipogenic differentiation, and osteogenic differentiation, which revealed that MSNs could serve as excellent probes and even drug carrier in the stem cell related applications [41].

The interactions of MSNs with the cells included the toxicity of the nanoparticles. Mou's group has also investigated the toxicity of different sizes and charges of MSNs. The positive charge of MSNs (50, 100 and 250 nm) and the negative charge of MSNs (100 nm) were synthesized and tested in lung epithelium BEAS-2B cells and human monocyte-derived macrophages. The results demonstrated that the amount of accumulation of MSNs inside the cells correlated to the toxicity, which appeared to be influenced by dosage and particle size. However, only positively charged with larger MSNs size ($>$100 nm) could sufficiently trigger the reactive oxygen species (ROS) and oxidative stress generation. Inflammatory genes were up-regulated by NF-κB and AP-1 in higher treatment doses ($\geq$500 $\mu g\,mL^{-1}$), which eventually led to autophagy-mediated necrotic cell death [50].

3.2 FMSNs for Labeling the Cell Surface Receptors

In situ labeling and tracking the cell surface receptors are important issues for early diagnosis, disease monitoring, and prevention. FMSNs provide a large area for the attachment of affinity ligands or antibodies to precisely recognize and bind with the receptors and antigens on the cell surfaces, which could be used for immunofluorescence and cell surface receptor labeling. As an example, anti-HER2/neu mAb conjugated to MSNs embedded green fluorescence dyes (Her-Dye@MSN) specifically recognized and targeted the BT-474 (HER2/neu over-expressing breast cancer cell) among other two cell lines NIH3T3 (mouse fibroblast cell) and MCF-7 (breast cancer cell), which could be potentially used as a cell imaging-based diagnostics as well as drug delivery platform [51]. Similarly, the peptide YSAYPDSVPMMSK (YSA) functionalized MSNs could bind specifically to EphA2 receptors on breast cancer cells [52]. Ri7 antibody conjugated to gadolinium-chelated Ri7-MSNs containing FITC through a polyethylene glycol (PEG) linker has been successfully used to label the brain microvessel endothelial cells to perform cell uptake studies and quantitative cellular assay [53]. In addition, ultra-bright fluorescent MSNs decorated with folic acid served as a fast and reliable diagnostics platform, which could distinguish between the cancerous and precancerous cervical epithelial cells from the normal cells. Therefore, the folic acid FMSNs could be used as alternative materials for prescreening of cervical cancers [54].

These FMSNs were also used for pathogenic bacteria sensing with minimal side effects. The hydrophobically modified sulforhodamine B and Cascade Blue hydrazide were incorporating with anti-gonococcus IgG antibody conjugated MSNs enabling stable and bright immunofluorescence labeling. This platform was introduced in rapid and ultrasensitive detection and specifically sorting of bacteria *Neisseria gonorrhoeae* using flow cytometry [55]. Furthermore, MSNs conjugated with FITC have been employed to track and visualize the MSN-vancomycin targeting and killing of the gram-positive bacteria such as *S. aureus* without harming the viability of the macrophage-like cells. The terminal of D-alanyl-D-alanine on the surfaces of gram-positive bacteria could specifically conjugate to vancomycin via multivalent hydrogen bonding interactions [56].

3.3 FMSNs for Intracellular Labeling, Tracking, and Sensing

Optical imaging and fluorescence probes hold the potential as a promising technique to investigate the spatiotemporal dynamics of ions in the living systems. Especially, monitoring the intracellular ion dynamics in real time requires high sensitivity and selectivity probes. FMSNs provide a platform for intracellular ions sensing via analyte-induced changes in photoluminescence. NaF is one of many ions, which are essential for cell living.

However, the excessive fluoride ions intake in cells leads to the imbalance of cell signaling components and could induce cell apoptosis. Therefore, quantitative detection of fluoride (particularly NaF) in the biological system is very important. For instance, incorporating with Ir-complex, MSNs could serve as a ratiometric pH sensing platform to detect the fluoride ion inside the cells. To fabricate F^- sensing platform, complex Ir2 (a blue-emitting Ir(III) complex) was covalently doped into MSNs via co-condensation. Subsequently, the complex Ir1′ (a yellow phosphorescent Ir(III)), which contained *tert*-butyldiphenylsilyl (TBDPS) moiety for F^- reaction site, were covalently grafted on the Ir2-MSNs. The Ir2 complex served as a reference, which was independent of the F^- concentration. Contrary, the presence of F^- ions triggered the specific Si–O cleavage reaction on the surfaces of MSNs, which induced the changes in fluorescence intensity of Ir1 complex. The Ir2-MSN-Ir1′ served as both time-resolved luminescence imaging and ratiometric sensing probe of intracellular F^-[57]. A similar design has been used for ratiometric ClO^- sensing. Two different phosphorescent iridium (III) complexes were incorporated onto solid MSNs inner core and grafted on the surfaces of MSNs, respectively. The inner solid core of MSNs acted as a reference whereas the outer mesoporous layer of the nanoparticles sensitively responded to the ClO^- concentration. The presence of ClO^- eliminated the isomerization of the aldoxime groups (C=N—OH) of phosphorescent iridium (III) complexes on the surfaces. Such core–shell MSNs allow ratiometric sensing and photoluminescence lifetime imaging to selectively and sensitively detect exogenous and endogenous ClO^-[58].

Beside the intracellular ion, the temperature is another important parameter for studying the cellular functions because it regulates the kinetics of almost all biochemical reactions and all the important cellular events, ranging from energy metabolism and gene expression to cell division. Dong et al. developed thermo-sensitive nanohybrid particles, which incorporated different materials such as polymer, MSNs, carbon dots, and dyes sensitive temperature to accurately measure the temperature at the single cell level and could be potentially used for biologically controlled thermal therapy. The temperature sensitive polymer, Poly-*N*-vinyl imidazole and 1-vinyl-2-(hydroxymethyl)-imidazole (UPVH), an upper critical solution temperature (UCST), were first synthesized (Fig. 4) and then were decorated on the surfaces of carbon dot-MSNs (core, CD-MSNs) as the thermo-sensitive shell generating thermo-sensitive nanohybrids (CD-MSN@UP38). The temperature changes could be accessed by fluorescence intensity and lifetime

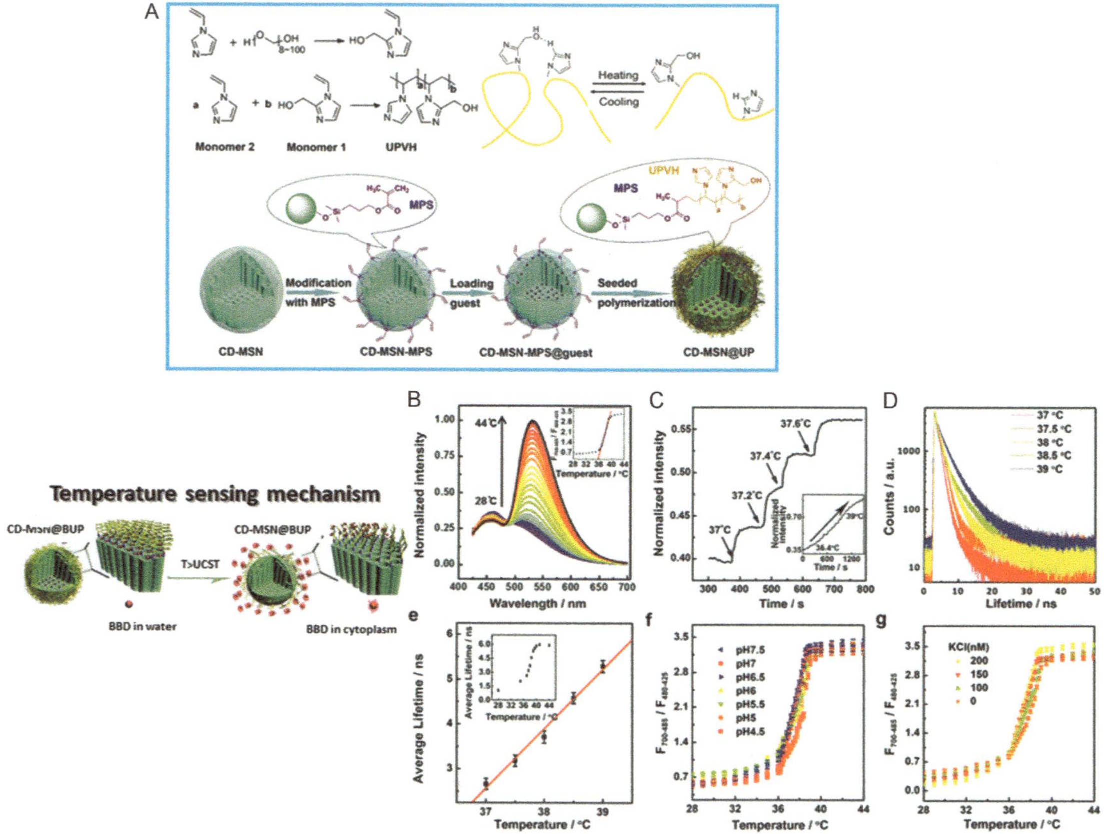

Fig. 4 (A) Schematic illustrating the fabrication of CD-MSN, CD-MSN-MPS, CD-MSN-MPS@guest, and CD-MSN@UP. Fluorescence properties of CD-MSN@BUP at different temperatures: (B) Fluorescence spectra of CD-MSN@BUP38 in a SH-SY5Y cell lysis solution with increasing the temperature from 28 to 44 °C. Excitation was set at 405 nm. Inset: Plot of $F_{700-485}/F_{480-425}$ as a function of temperature (error bars: SD, $n=3$). (C) Thermo-responsive efficiency of CD-MSN@BUP38 measured via fluorescence intensity of BBD. (D) Fluorescence emission decays of CD-MSN@BUP38 at 530 nm emission with temperature variation at 35, 37, 37.5, 38, and 38.6 °C in a SH-SY5Y cell lysis solution. (E) Relationship between the average lifetime of CD-MSN@BUP38 and temperature (error bars: SD, $n=3$). (F and G) Plots of $F_{700-485}/F_{480-425}$ obtained from CD-MSN@BUP38 as a function of temperature in (F) different pH values from 4.75 to 7.5, and (G) different ion strength environments with KCl concentrations of 0, 100, 150, 200 nM (error bars: SD, $n=3$) [59].

where polarity-sensitive dye 4-benzylamino-7-nitro-2,1,3-benzoxadiazole (BBD) was loaded into the pores CD-MSNs (CD-MSN@BUP38). This novel thermometer CD-MSN@BUP38 exhibited rapid temperature responses toward variations down to 0.21 °C which allowed accurately mapping and sensing of intracellular temperature with high spatial resolution using fluorescence lifetime imaging microscopy (FLIM). The fluorescence lifetime-based sensing method is independent of any experimental fluctuations, including the concentration of fluorescent probes or the excitation source [59].

The diverse intracellular pH (pHi) distribution is important to maintain various biological reactions in different organelles. Due to the heterogeneity of pH in the cells, the pH microenvironment of nanoparticles could be used as an indicator of intracellular location. Chen et al. successfully mapped out the positions of MSNs by monitoring the local pH surrounding the MSNs using single particle tracking. This result revealed the interaction between nanoparticles and membrane cells was mediated by charge. In this design, the MSNs ratiometric pH sensor was fabricated by encapsulating FITC as a pH-sensitive dye and RITC as a reference inside MSNs [60]. On the other hand, Chiu et al. also took the advantage of the intracellular pH to trigger release chromobodies from their MSN vesicles. In their study, the large pores of MSNs were functionalized with nitrilotriacetic acid–metal ion complexes such as Fe^{2+}, Co^{2+}, Ni^{2+}, Cu^{2+}, Zn^{2+}, and Ca^{2+}, which were used to efficiently carry and release His6-tagged chromobodies into the cells. In order to observe the binding affinity and pH-trigger release of the complex, the in vitro colorimetric measurements have been performed. Furthermore, 3D Structure Illumination Microscopy technique has been used to determine the locations of the chromo bodies in real time. It was found that the acidic environment of endosome triggered the release of chromobodies from MSNs, which successfully bound to the antigen in the cytosol and provided specifically antigen visualization [61]. Rhodamine B (RhB) molecules with high quantum yield were utilized as cargo inside the hybrid copolymer/MSNs to examine the redox-responsive control release due to the cleavage of sulfide bonds. Moreover, decorating the hybrid MSNs with fluorescent pH sensing monomer (1,8-naphthalimide) enabled them to monitor the pH in cells and tissues [62]. Furthermore, Mou and co-worker synthesized the large pore MSNs containing a near-infrared (NIR) fluorophore (Atto-647) to transport the DOX into the Hep-G2 cells. To provide the pH stimuli site cleavage, pH-sensitive linkers were grafted onto the walls of nanopores via hydrazone bond. These nanopore modifications facilitated the loading of DOX-hydrazone onto MSN pores and formed the double hydrazine link. Under the acidic conditions in endosomes and lysosomes, the unstable double hydrazine link was cleaved and caused the release of DOX into the cells. Combination strategy of encapsulation and pH-sensitive functionalization enhanced the localization ability of MSNs to release their payloads and prevented the premature leakage of the drug. Moreover, incorporating Atto-647 into MSN matrix allowed the delivery and release process to be simultaneously monitored under NIR light [63].

3.4 FMSNs for Monitoring Control Release Process

Together with their function as a diagnostic tool, fluorophores decorated on the MSNs could subsequently provide an additional important function, as a controller for the release of encapsulated drugs. Various inorganics fluorophores, such as Au, QDs, CD, UCNP, were also conjugated onto the surfaces of MSNs as a gatekeeper, to control the release of drugs, with either intracellular or external stimuli trigger (such as pH, enzyme, reduction, light irradiation). For example, fluorescein molecules were covalently linked inside lipid bilayer dipalmitoyl moieties supporting phosphorylated lipid MSNs. Incorporating of fluorescein molecules in this system has two purposes. The first one was to monitor their biocompatibility and cell internalization of the nanocarriers in both normal and cancer cell lines using fluorescence spectroscopy and flow cytometry. The second one was to control the release of fluorescein cargo in the presence of disulfide reducing conditions, which could be monitored optically [64]. A similar strategy was used by covalently anchored fluorescent dye (Texas Red™) to the MSNs and entrapped the cadmium sulfides (CdS) QDs inside their mesopores. The QDs release was observed by monitoring the FRET intensity between the two fluorophores [65]. It is worth to mention that the loading of enhanced green fluorescent protein (eGFP) modified bovine serum albumin (BSA) in gold nanoparticles-functionalized MSNs was used to carry and deliver the non-denatured proteins and plasmid DNA successfully to plant tissues [66].

Moreover, for monitoring the anticancer drugs release in real time, the pores of MSNs were grafted with paracyclophane fluorophores (CF), which exhibited large two-photon absorption. The pseudo-rotaxane was decorated onto pores of MSNs loaded with anticancer drugs via azobenzene stalk and a β-cyclodextrin moiety (Fig. 5). The FMSNs could be used to image the MCF-7 breast cancer cells, at low power of two-photon irradiation, whereas the high power irradiation has been utilized to trigger pore opening via FRET, which led to the release of anticancer drugs [67].

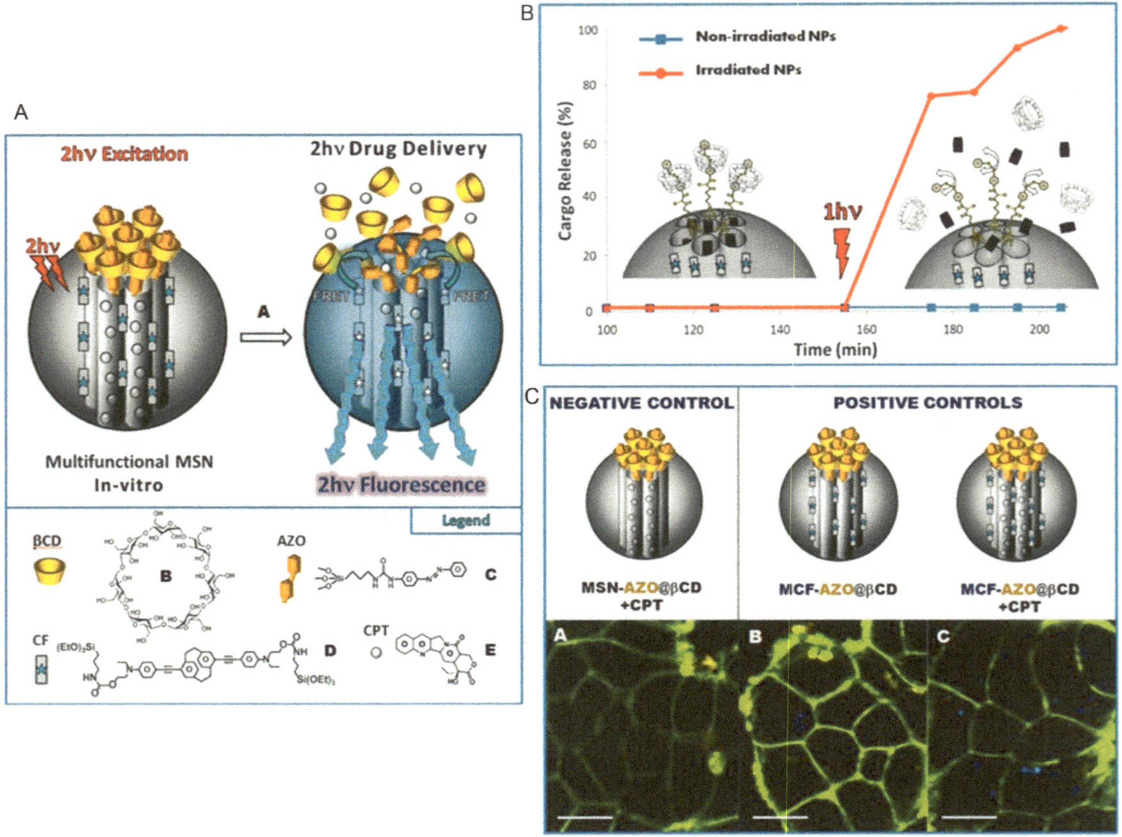

Fig. 5 (a) Schematic representation of the two-photon-triggered multifunctional nanovalve. The two-photon irradiation (**A**) triggers the release of the camptothecin anticancer drugs (**E**), via the photo-isomerization of azobenzene moieties (**C**) grafted on the surface of CF fluorophore (**D**)-modified MSN, and linked with β-cyclodextrin (**B**). (b) One-photon triggered propidium iodide release in water, the comparison between irradiated and non-irradiated nanovalves MCF-AZO@βCD NPs at 365 nm (2–3 wt% or dye released). (c) Two-photon imaging of MSN-AZO@βCD loaded with camptothecin (A), MCF-AZO@βCD with and without camptothecin (B–C), confirming the cellular uptake of the nanovalves, as well as their potential for two-photon intracellular tracking. Scale bar of 10 μm [67].

4. BIODISTRIBUTION AND FATE OF FUNCTIONALIZED MSNs

For the in vivo studies, various biological compositions could attach to the surfaces of MSNs, which can eventually lead to the clearance by the reticuloendothelial system (RES). For example, within a few minutes in the human serum, nanoparticles have been already crowned with the plasma corona, which is known as "opsonization." The opsonization process is a dynamic process which can be divided into two categories: (i) the "soft-corona" consisting loosely bound proteins, which can be replaced easily and (ii) the "hard corona" with tightly bound proteins, which cannot be replaced by other molecules. Landry's group recently utilized STORM for protein mapping on the surfaces of MSNs. In this study, two different pore sizes of MSNs (3 and 6 nm pore MSNs) were used to demonstrate the penetration depth of different size of proteins (17, 66, and 138 kDa) into MSN pores. The results suggested that the smaller size of protein could penetrate deeper into the pores. It was also found that small proteins easily adsorbed and resided inside the large pores, while large proteins were only bound to the surfaces of the corona. Therefore, the majority in the corona composition were the low molecular weight (<50 kDa) proteins [68]. Moreover, Lorenzo's group unveiled that the surface chemistry led to the diversity of the corona. The heterogeneity of the corona dramatically increased over time and was enhanced by particle degradability [69]. Hence, the presence of proteins and other biomolecules in the bloodstream could impact on the FMSN's biological fate.

The opsonized nanoparticles could easily be captured by RES or MPS, resulting in a quick clearance of nanoparticles from the blood. Furthermore, if the nanoparticles could escape from the RES or MPS, the corona formation of nanoparticle also would affect the cell targeting process. Therefore, a more detailed understanding of nanoparticle biodistribution is needed. It was reported that physicochemical properties of nanoparticles such as shape, size, and surface functionalization strongly influenced the blood retention of MSNs. Huang et al. utilized two different shapes of MSNs conjugated FITC (with aspect ratios of 1.5 and 5) to investigate the influence of the shape to the nanoparticle behavior in the biodistribution and clearance and to the toxicity. The results indicated that intravenously administrated MSNs were mainly deposited in the liver, spleen, and lung (>80%). Moreover, the particle shapes significantly influenced their in vivo behaviors, both the clearance time and biodistribution. It was observed the short rod shape easily deposited inside the liver, whereas the long-rod shape stayed in the spleen. However, after PEGylation, the biodistribution of both MSNs shifted to the lung. For the clearance time, the short-rod MSN had a more rapid clearance rate than long-rod MSNs in both excretion routes, urine and feces. Furthermore, by using the intravital multiphoton microscopy Cheng et al. tracked the dynamics of hepatic metabolism in real time for FMSNs with different surface modification during the biodistribution in mice with subcellular resolution. The result reveals the uptakes of different charges of MSNs were contrary to each other between hepatocytes and Kupffer cells in liver sinusoids (Fig. 6). The FMSNs with positive charges were significantly uptaken by hepatocytes while negatively charged FMSNs were captured by Kupffer cells in liver sinusoids. The positively charged MSNs, which reside in hepatocytes, potentially could be removed via hepatobiliary excretion.

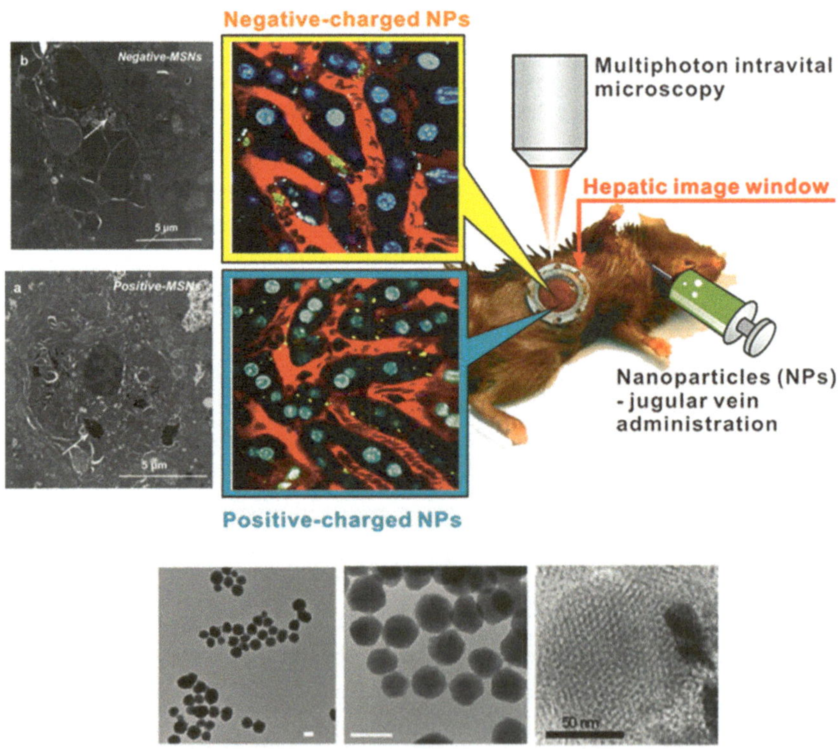

Fig. 6 Multiphoton intravital microscopy visualizes dynamics of the sub-hepatic distribution of different surface-charged nanoparticles [70].

However, the negatively charged MSNs accumulated in Kupffer cells indicating significant hepatotoxicity [70].

In order to minimize the nanoparticle opsonization and to slow down the clearance of nanoparticles, several strategies have been developed. One common strategy is to minimize the particle aggregation by maintaining the steric stabilization or to increase electrostatic repulsion between particles. Lee et al. [34] encapsulate the indocyanine green (ICG) into MCM-41 modified with trimethylammonium (TA) via electrostatic interactions to protect them from self-quenched, degradation, and prevent the possible of variation of optical properties caused by different physiological conditions [34]. The encapsulation of ICG molecules into MSNs enhanced the number of photon count as compared with free ICG solutions. Due to their brightness, they were able to monitor the distribution of MSNs in the anesthetized rat. Moreover, their ability to visualize the MSNs, which were mostly accumulated in the organ such as liver, revealed that ICG-MSN could be a promising contrast agent for deep tissue imaging.

5. FMSNs AS A PROBE FOR IN VIVO CELL TRACKER

The ability to noninvasively monitor cell trafficking in vivo is an emerging requirement for the cellular therapeutic strategies. For instance, FITC-conjugated MSNs have been employed to track and could be efficiently used to label the iPSCs and simultaneously carry the hepatocyte nuclear factor 3β (HNF3β) plasmid DNA (pDNA) into iPSCs resulting in enhancement of differentiation into hepatocyte-like cells [71].

On the other hand, Heidegger et al. developed fluorescent dye-conjugated MSNs for immune cells tracking. The FITC was conjugated on the amine-modified MSNs to serve as a tracker and polyvinyl pyridine (PVP) was subsequently decorated on the surfaces to maintain the colloidal stability. The results indicated that internalization of MSNs was concentration-dependent in isolated splenocytes from the mouse. Moreover, among all other cells, dendritic cells, which are highly specialized in antigen uptake, exhibited the most efficient of MSNs uptake. Conversely, T and B cells did not recognize the MSNs showing only a few of fluorescence signal. Furthermore, speckled pattern of fluorescent signals within dendritic cells was observed, suggesting an uptake of FITC-labeled MSNs into organelles, but not into the cytosol or nucleus [31].

For cancer cell tracking and therapeutic targeting, DOX-loaded MSNs served both as cancer metastasis chemopreventive agent delivery and simultaneously self-monitoring for targeting the circulating tumor cells (CTCs). The incorporation with EpCAM and CD44 aptamers on the surfaces MSNs enabled the particles to specifically recognize, bind, efficiently deliver drugs into colorectal cancer cells, and resulting in down-regulated effect to the activity of the captured colorectal CTCs through disturbing the EpCAM and CD44 expression on the colorectal CTCs cell surfaces, inhibiting the hetero-adhesion between the cancer cells and endothelial cells, moderately controlling cell proliferation, and restraining colorectal CTCs cell transmigration [72].

6. FMSNs FOR TUMOR IMAGING AND THE THERANOSTICS APPLICATIONS

The potential uses of MSNs for the tumor therapy have been extended by incorporating more than one strategy of treatment in the same system. For instance, high capacity loading (1.94×10^4 molecules/particle) of photosensitizer protoporphyrin IX (PpIX) inside the wall of the pores of MSNs has been reported by Mou's group for photodynamic therapy (PDT). In this strategy, the walls of the pores in MSNs were firstly functionalized with amine groups to allow the conjugation of PpIX via a covalent bond. The encapsulation of PpIX into the nanopores provided site isolation to prevent the PpIX self-quenching and the silica matrix protected them from the harsh environmental condition. Under 514 nm light excitation, the PpIX-modified MSNs (PpIX-MSNs) generated the singlet oxygen, which led to the phototoxicity effect. To evaluate in vitro PDT effect, the complex PpIX-MSNs were incubated in HeLa cells. The results suggested the PpIX-MSNs were efficiently internalized and exhibited a high efficacy of PDT effect, which highly depended on the dosage accumulation and time irradiations. The phototoxicity effect successfully induced the cell death through both necrosis and apoptosis [73]. Similarly, oxygen sensing probe was covalently conjugated into silica to achieve the high efficacy of PDT effect in breast cancer cells. The PDT effect can be achieved by simply increasing the energy of irradiation to switch the function of Pd–porphyrin (PdTPP) from oxygen sensing probe to photosensitizer. Integrating PdTPP into MSNs facilitated the cell internalization and prevented the probe leakage. Moreover, this system showed negligible toxicity during the transportation and before the irradiation. However, when the accumulation of dosage inside the cells was higher than 25 μg/mL and the irradiation energy was larger than $2.4\,\mathrm{J\,cm^{-2}}$, it triggered the cell death via necrosis [74]. The theranostics applications of PdTPP were explored where PdTPP served as both in vivo oxygen sensing probe and photosensitizer by covalently loading them into the pores of MSNs conjugated NIR-ATTO647N. The surfaces of PdTPP-complex MSNs were conjugated with cRGDyK peptides for specifically targeting the cancer cells with overexpressed αvβ3 integrins. Such a design allowed monitoring the process of targeting and producing PDT effect in the targeted cells with a very minimum side effect [75].

Integrating multiple imaging probes into one single platform is a promising strategy in the tumor imaging. Wu et al. were the first to report the applications of MSNs for bimodal imaging in the animal model. The biomodal probes consisting of $Fe_3O_4@SiO_2$ and MSN-FITC enabled direct monitoring of the behavior of nanoparticles in vivo including their cytotoxicity, cellular internalization, accumulation and biodistribution in the animal model. To test the efficiency of cell labeling, these hybrid MSNs were incubated in both rat bone marrow stromal cells (rMSCs) and NIH 3T3 fibroblast cells. The data suggested the hybrid nanoparticles allowed high-efficiency labeling of rMSCs. Moreover, the in vitro T2-weighted MR results demonstrated that most likely the hybrid MSNs resided in the liver and spleen tissue more than in kidney and only a few of them were internalized by macrophages [76]. Chen et al. incorporated the copper-64 (^{64}Cu, $t_{1/2}=12.7$ h) and 800CW (a near-infrared fluorescence dye) onto MSNs to successfully image tumor vasculature in vivo using a multimodal approach, which employed both positron emission tomography (PET) imaging and NIR imaging techniques. To specifically target the tumor, TRC105 (a human/murine chimeric IgG1 monoclonal antibody) was grafted onto the surfaces of MSNs against CD10, a vascular-specific marker for tumor angiogenesis. The results demonstrated that there was significant improvement in both PET and fluorescence imaging resolution in vivo compared to non-targeted controls [77]. Moreover, the in vivo applications of MSNs are not limit to bioimaging because optical probe is often co-delivery with bioactive molecules, resulting in a theranostic platform. Highly fluorescent molecules, either fluorescent dyes with no therapeutic effect or fluorescent drugs, are often chosen as a cargo. For instance, rattle-structured with a controllable cavity containing Gd-upconversion nanoparticles acting as a core and a porous silica layer as the outer shell has been synthesized by using a "surface-protected etching" strategy (UCSNs). UCSNs served as a magnetic/luminescent dual-mode imaging probe for identifying and recognizing the location of the tumor in vivo while the hollow cavity and porous shell structure could store, protect, and deliver a sufficient dosage of cisplatin for chemo and radiotherapy [78]. Similarly, in order to obtain imaging-guided therapy of cancer, two-dimensional magnetic WS2@Fe3O_4 nanocomposite with mesoporous silica coating has been developed. The Cy5. 5-embedded MSNs coated with WS2 nanosheets have been synthesized as the core and the core was shielded with iron oxide layer via self-assembly to form core/shell structure. The strong NIR emission and X-ray absorption were generated from the core, making these nanoparticles suitable for photothermal and CT probes. Moreover, these nanoparticles were biocompatible for various types of cells. For the imaging applications, the nanoparticles were administrated via intravenous injection enabling the localization of tumors with great fluorescence, MR, and CT signals enhancement. This system has also been used for doxorubicin (DOX) delivery resulting in completely inhibition of tumor growth in the mice under laser irradiation [79]. Furthermore, FMSNs have been used as the endoscopic contrast agents for diagnostics of premalignant colonic lesions. Lo's group utilized FITC doped-MSNs coated with PEG and conjugated with UEA1 for early diagnostics of polyps and nascent colorectal cancers. The FITC doped-MSNs allowed sensitively monitoring in vivo targeting of α-L-fucose receptors by in vivo endoscopic and microscopic imaging [80].

7. CONCLUSIONS AND FUTURE PERSPECTIVE

In summary, the unique properties of MSNs, such as tunable pore sizes, modifiable morphology (controllable particle shape and size), hydrophilic nature, high loading capacity, and suitability surface modification, make them very useful for drug delivery system. Because MSNs do not emit light, many strategies have been developed to engineer fluorescent MSNs for visualization in cells and animals. Even though there is enormous progress in the development of functional MSNs for various imaging applications, clinical translation still requires a lot of research efforts.

Nevertheless, several key opportunities and/or challenges still exist in this area. First, the standardizations are required to achieve reproducibility in the synthesis of FMSNs with controllable shape and size, good stability and dispersibility, and multifunctionalization surfaces for clinical applications. Second, further research efforts for surface engineering of MSNs are needed to improve the stability of MSNs in the bloodstream and to suppress nonspecific MSNs uptake by the RES. Furthermore, for translational studies, the interaction between MSNs with complex biological system, especially for large animals, needs to be addressed. Third, the development of FMSNs applications for personalized medicine using stem cells and activation of the patient's immune system may be the future directions of nanomedicine. In the end, we anticipate that such challenges or opportunities on the developing FMSNs could potentially be overcome through the interdisciplinary conversations between researchers in the fields of biology, materials science, nanotechnology, chemistry, optical imaging, computation and modeling, and clinical medicine. Meanwhile, the development of MSNs should always keep eyes on the emerging tools such as imaging technologies, new nanomaterials, and new drugs and antibodies.

REFERENCES

[1] S.W. Hell, J. Wichmann, Breaking the diffraction resolution limit by stimulated emission: stimulated-emission-depletion fluorescence microscopy, Opt. Lett. 19 (11) (1994) 780–782.
[2] C. Eggeling, et al., Direct observation of the nanoscale dynamics of membrane lipids in a living cell, Nature 457 (7233) (2009) 1159–1162.
[3] L. Schermelleh, et al., Subdiffraction multicolor imaging of the nuclear periphery with 3D structured illumination microscopy, Science 320 (5881) (2008) 1332–1336.
[4] E. Betzig, et al., Imaging intracellular fluorescent proteins at nanometer resolution, Science 313 (5793) (2006) 1642.
[5] S. Manley, et al., High-density mapping of single-molecule trajectories with photoactivated localization microscopy, Nat. Methods 5 (2) (2008) 155–157.
[6] B. Huang, et al., Three-dimensional super-resolution imaging by stochastic optical reconstruction microscopy, Science 319 (5864) (2008) 810–813.
[7] M.J. Rust, M. Bates, X. Zhuang, Sub-diffraction-limit imaging by stochastic optical reconstruction microscopy (STORM), Nat. Methods 3 (10) (2006) 793–796.
[8] S. Mura, J. Nicolas, P. Couvreur, Stimuli-responsive nanocarriers for drug delivery, Nat. Mater. 12 (11) (2013) 991–1003.
[9] C. Giménez, et al., Gated mesoporous silica nanoparticles for the controlled delivery of drugs in cancer cells, Langmuir 31 (12) (2015) 3753–3762.
[10] C.-Y. Lai, et al., A mesoporous silica nanosphere-based carrier system with chemically removable CdS nanoparticle caps for stimuli-responsive controlled release of neurotransmitters and drug molecules, J. Am. Chem. Soc. 125 (15) (2003) 4451–4459.
[11] Q. Cai, et al., Dilute solution routes to various controllable morphologies of MCM-41 silica with a basic medium, Chem. Mater. 13 (2) (2001) 258–263.
[12] C.E. Fowler, et al., Nanoscale materials with mesostructured interiors, Adv. Mater. 13 (9) (2001) 649–652.
[13] R.I. Nooney, et al., Synthesis of nanoscale mesoporous silica spheres with controlled particle size, Chem. Mater. 14 (11) (2002) 4721–4728.
[14] C.T. Kresge, et al., Ordered mesoporous molecular sieves synthesized by a liquid-crystal template mechanism, Nature 359 (1992) 710.
[15] H. Frank, et al., Silica-based mesoporous organic–inorganic hybrid materials, Angew. Chem. Int. Ed. 45 (20) (2006) 3216–3251.
[16] Y.-S. Lin, et al., Well-ordered mesoporous silica nanoparticles as cell markers, Chem. Mater. 17 (18) (2005) 4570–4573.

[17] M. Shigeyuki, O. Yuya, I. Hiroaki, Incorporation of dyes into silica–surfactant mesostructured nanoparticles as a nanoscale host material for organic molecules, Chem. Lett. 35 (8) (2006) 880–881.

[18] S. Igor, Y.Y. Kievsky, J.M. Kaszpurenko, Self-assembly of ultrabright fluorescent silica particles, Small 3 (3) (2007) 419–423.

[19] T.-W. Kim, P.-W. Chung, V.S.Y. Lin, Facile synthesis of monodisperse spherical MCM-48 mesoporous silica nanoparticles with controlled particle size, Chem. Mater. 22 (17) (2010) 5093–5104.

[20] A. Liu, et al., Fluorescent hybrid with electron acceptor methylene viologen units inside the pore walls of mesoporous MCM-48 silica, Langmuir 26 (5) (2010) 3555–3561.

[21] H. Suk Ho, K. Hyunjin, C. Yongdoo, Indocyanine green-loaded hollow mesoporous silica nanoparticles as an activatable theranostic agent, Nanotechnology 28 (18) (2017) 185102.

[22] Y.-S. Lin, et al., Synthesis of hollow silica nanospheres with a microemulsion as the template, Chem. Commun. 24 (2009) 3542–3544.

[23] X.-J. Wu, D. Xu, Formation of yolk/SiO2 shell structures using surfactant mixtures as template, J. Am. Chem. Soc. 131 (8) (2009) 2774–2775.

[24] Y. Pan, et al., Gd-based upconversion nanocarriers with yolk-shell structure for dual-modal imaging and enhanced chemotherapy to overcome multidrug resistance in breast cancer, Nanoscale 8 (2) (2016) 878–888.

[25] L. Zhao, et al., Yolk–shell upconversion nanocomposites for LRET sensing of cysteine/homocysteine, ACS Appl. Mater. Interfaces 6 (14) (2014) 11190–11197.

[26] Z. Teng, et al., Mesoporous silica hollow spheres with ordered radial mesochannels by a spontaneous self-transformation approach, Chem. Mater. 25 (1) (2013) 98–105.

[27] Z. Tierui, et al., Formation of hollow silica colloids through a spontaneous dissolution–regrowth process, Angew. Chem. Int. Ed. 47 (31) (2008) 5806–5811.

[28] J. Lei, L. Wang, J. Zhang, Superbright multifluorescent core – shell mesoporous nanospheres as trackable transport carrier for drug, ACS Nano 5 (5) (2011) 3447–3455.

[29] J. Kim, et al., Magnetic fluorescent delivery vehicle using uniform mesoporous silica spheres embedded with monodisperse magnetic and semiconductor nanocrystals, J. Am. Chem. Soc. 128 (3) (2006) 688–689.

[30] J. Ciccione, et al., Unambiguous and controlled one-pot synthesis of multifunctional silica nanoparticles, Chem. Mater. 28 (3) (2016) 885–889.

[31] S. Heidegger, et al., Immune response to functionalized mesoporous silica nanoparticles for targeted drug delivery, Nanoscale 8 (2) (2016) 938–948.

[32] B. Ahn, et al., Mesoporous silica nanoparticle-based cisplatin prodrug delivery and anti-cancer effect under reductive cellular environment, J. Mater. Chem. B 1 (22) (2013) 2829–2836.

[33] X. Huang, et al., The shape effect of mesoporous silica nanoparticles on biodistribution, clearance, and biocompatibility in vivo, ACS Nano 5 (7) (2011) 5390–5399.

[34] L. Chia-Hung, et al., Near-infrared mesoporous silica nanoparticles for optical imaging: characterization and in vivo biodistribution, Adv. Funct. Mater. 19 (2) (2009) 215–222.

[35] S. Sreejith, X. Ma, Y. Zhao, Graphene oxide wrapping on squaraine-loaded mesoporous silica nanoparticles for bioimaging, J. Am. Chem. Soc. 134 (42) (2012) 17346–17349.

[36] T.R. Sathe, A. Agrawal, S. Nie, Mesoporous silica beads embedded with semiconductor quantum dots and iron oxide nanocrystals: dual-function microcarriers for optical encoding and magnetic separation, Anal. Chem. 78 (16) (2006) 5627–5632.

[37] L. Juying, et al., Carbon dot-incorporated PMO nanoparticles as versatile platforms for the design of ratiometric sensors, multichannel traceable drug delivery vehicles, and efficient photocatalysts, Adv. Opt. Mater. 3 (1) (2015) 57–63.

[38] S. Pandey, et al., Synthesis of mesoporous silica oxide/C-dot complex (meso-SiO2/C-dots) using pyrolysed rice husk and its application in bioimaging, RSC Adv. 4 (3) (2014) 1174–1179.
[39] L. Jianan, et al., Controlled synthesis of uniform and monodisperse upconversion core/mesoporous silica Shell nanocomposites for bimodal imaging, Chem. Eur. J. 18 (8) (2012) 2335–2341.
[40] I. Slowing, B.G. Trewyn, V.S.Y. Lin, Effect of surface functionalization of MCM-41-type mesoporous silica nanoparticles on the endocytosis by human cancer cells, J. Am. Chem. Soc. 128 (46) (2006) 14792–14793.
[41] T.-H. Chung, et al., The effect of surface charge on the uptake and biological function of mesoporous silica nanoparticles in 3T3-L1 cells and human mesenchymal stem cells, Biomaterials 28 (19) (2007) 2959–2966.
[42] L. Fang, et al., Size effect on cell uptake in well-suspended, uniform mesoporous silica nanoparticles, Small 5 (12) (2009) 1408–1413.
[43] W. Jiang, et al., Nanoparticle-mediated cellular response is size-dependent, Nat. Nanotechnol. 3 (2008) 145.
[44] B.D. Chithrani, A.A. Ghazani, W.C.W. Chan, Determining the size and shape dependence of gold nanoparticle uptake into mammalian cells, Nano Lett. 6 (4) (2006) 662–668.
[45] M. Aoyama, et al., Intracellular trafficking of particles inside endosomal vesicles is regulated by particle size, J. Control. Release 260 (2017) 183–193.
[46] Y.-P. Chen, et al., Surface charge effect in intracellular localization of mesoporous silica nanoparticles as probed by fluorescent ratiometric pH imaging, RSC Adv. 2 (3) (2012) 968–973.
[47] M. Liong, et al., Multifunctional inorganic nanoparticles for imaging, targeting, and drug delivery, ACS Nano 2 (5) (2008) 889–896.
[48] C.-H. Wu, et al., Enhanced non-endocytotic uptake of mesoporous silica nanoparticles by shortening the peptide transporter arginine side chain, ACS Appl. Mater. Interfaces 5 (23) (2013) 12244–12248.
[49] Y. Jian, et al., Membrane fusion mediated intracellular delivery of lipid bilayer coated mesoporous silica nanoparticles, Adv. Healthc. Mater. 6 (20) (2017) 1700759.
[50] C.-C. Chou, et al., Molecular elucidation of biological response to mesoporous silica nanoparticles in vitro and in vivo, ACS Appl. Mater. Interfaces 9 (27) (2017) 22235–22251.
[51] C.-P. Tsai, et al., Monoclonal antibody-functionalized mesoporous silica nanoparticles (MSN) for selective targeting breast cancer cells, J. Mater. Chem. 19 (32) (2009) 5737–5743.
[52] Z. Liu, et al., YSA-conjugated mesoporous silica nanoparticles effectively target EphA2-overexpressing breast cancer cells, Cancer Chemother. Pharmacol. 81 (4) (2018) 687–695.
[53] M. Bouchoucha, et al., Antibody-conjugated mesoporous silica nanoparticles for brain microvessel endothelial cell targeting, J. Mater. Chem. B 5 (37) (2017) 7721–7735.
[54] S. Palantavida, et al., Ultrabright fluorescent mesoporous silica nanoparticles for prescreening of cervical cancer, Nanomed. Nanotechnol. Biol. Med. 9 (8) (2013) 1255–1262.
[55] M.T. Hurley, et al., Synthesis, characterization, and application of antibody functionalized fluorescent silica nanoparticles, Adv. Funct. Mater. 23 (26) (2013) 3335–3343.
[56] G. Qi, et al., Vancomycin-modified mesoporous silica nanoparticles for selective recognition and killing of pathogenic gram-positive bacteria over macrophage-like cells, ACS Appl. Mater. Interfaces 5 (21) (2013) 10874–10881.
[57] S. Liu, et al., Reaction-based phosphorescent nanosensor for ratiometric and time-resolved luminescence imaging of fluoride in live cells, Chem. Commun. 51 (64) (2015) 12839–12842.

[58] K.Y. Zhang, et al., Core-shell structured phosphorescent nanoparticles for detection of exogenous and endogenous hypochlorite in live cells via ratiometric imaging and photoluminescence lifetime imaging microscopy, Chem. Sci. 6 (1) (2015) 301–307.
[59] F. Dong, et al., An engineered thermo-sensitive nanohybrid particle for accurate temperature sensing at the single-cell level and biologically controlled thermal therapy, J. Mater. Chem. B 4 (47) (2016) 7681–7688.
[60] C.-J. Tsou, et al., Local pH tracking in living cells, Nanoscale 7 (9) (2015) 4217–4225.
[61] H.-Y. Chiu, et al., Intracellular chromobody delivery by mesoporous silica nanoparticles for antigen targeting and visualization in real time, Sci. Rep. 6 (2016) 25019.
[62] X. Wan, D. Wang, S. Liu, Fluorescent pH-sensing organic/inorganic hybrid mesoporous silica nanoparticles with tunable redox-responsive release capability, Langmuir 26 (19) (2010) 15574–15579.
[63] L. Chia-Hung, et al., Intracellular pH-responsive mesoporous silica nanoparticles for the controlled release of anticancer chemotherapeutics, Angew. Chem. Int. Ed. 49 (44) (2010) 8214–8219.
[64] R.A. Roggers, V.S.Y. Lin, B.G. Trewyn, Chemically reducible lipid bilayer coated mesoporous silica nanoparticles demonstrating controlled release and HeLa and normal mouse liver cell biocompatibility and cellular internalization, Mol. Pharm. 9 (9) (2012) 2770–2777.
[65] C.Y. Lai, et al., Reversible binding and fluorescence energy transfer between surface-derivatized CdS nanoparticles and multi-functionalized fluorescent mesoporous silica nanospheres, in: R. Xu et al. (Ed.), Studies in Surface Science and Catalysis, Elsevier, 2007, pp. 1827–1835.
[66] M.-O. Susana, et al., Gold functionalized mesoporous silica nanoparticle mediated protein and DNA codelivery to plant cells via the biolistic method, Adv. Funct. Mater. 22 (17) (2012) 3576–3582.
[67] C. Jonas, et al., Two-photon-triggered drug delivery via fluorescent nanovalves, Small 10 (9) (2014) 1752–1755.
[68] A.M. Clemments, P. Botella, C.C. Landry, Spatial mapping of protein adsorption on mesoporous silica nanoparticles by stochastic optical reconstruction microscopy, J. Am. Chem. Soc. 139 (11) (2017) 3978–3981.
[69] F.-G. Natalia, et al., Super-resolution microscopy unveils dynamic heterogeneities in nanoparticle protein corona, Small 13 (41) (2017) 1701631.
[70] S.-H. Cheng, et al., Visualizing dynamics of sub-hepatic distribution of nanoparticles using intravital multiphoton fluorescence microscopy, ACS Nano 6 (5) (2012) 4122–4131.
[71] W. Chen, et al., Nonviral cell labeling and differentiation agent for induced pluripotent stem cells based on mesoporous silica nanoparticles, ACS Nano 7 (10) (2013) 8423–8440.
[72] Y. Gao, et al., A novel nanomissile targeting two biomarkers and accurately bombing CTCs with doxorubicin, Nanoscale 9 (17) (2017) 5624–5640.
[73] T. Hsiung-Lin, et al., In vitro studies of functionalized mesoporous silica nanoparticles for photodynamic therapy, Adv. Mater. 21 (2) (2009) 172–177.
[74] S.-H. Cheng, et al., Mesoporous silica nanoparticles functionalized with an oxygen-sensing probe for cell photodynamic therapy: potential cancer theranostics, J. Mater. Chem. 19 (9) (2009) 1252–1257.
[75] S.-H. Cheng, et al., Tri-functionalization of mesoporous silica nanoparticles for comprehensive cancer theranostics—the trio of imaging, targeting and therapy, J. Mater. Chem. 20 (29) (2010) 6149–6157.
[76] W. Si-Han, et al., Multifunctional mesoporous silica nanoparticles for intracellular labeling and animal magnetic resonance imaging studies, Chembiochem 9 (1) (2008) 53–57.

[77] F. Chen, et al., In vivo tumor vasculature targeted PET/NIRF imaging with TRC105 (fab)-conjugated, dual-labeled mesoporous silica nanoparticles, Mol. Pharm 11 (11) (2014) 4007–4014.
[78] W. Fan, et al., Rattle-structured multifunctional nanotheranostics for synergetic chemo-/radiotherapy and simultaneous magnetic/luminescent dual-mode imaging, J. Am. Chem. Soc. 135 (17) (2013) 6494–6503.
[79] G. Yang, et al., Two-dimensional magnetic WS2@Fe3O4 nanocomposite with mesoporous silica coating for drug delivery and imaging-guided therapy of cancer, Biomaterials 60 (2015) 62–71.
[80] N.-T. Chen, et al., Lectin-functionalized mesoporous silica nanoparticles for endoscopic detection of premalignant colonic lesions, Nanomed. Nanotechnol. Biol. Med. 13 (6) (2017) 1941–1952.

CHAPTER SEVEN

Biodistribution and Excretion of Intravenously Injected Mesoporous Silica Nanoparticles: Implications for Drug Delivery Efficiency and Safety

Mika Lindén[1]
Department of Inorganic Chemistry II, Ulm University, Ulm, Germany
[1]Corresponding author: e-mail address: mlinden@uni-ulm.de

Contents

Abstract

Mesoporous silica nanoparticles (MSNs) are currently attracting a high interest for use as drug carriers *in vivo*. To date only data on the biodistribution in small animals are available. As any nanoparticle system, the MSNs typically accumulate in the RES organs lung, liver, and spleen upon intravenous (i.v.) administration. However, the literature data are partly inconclusive, which can be connected to the wide variability of the experimental designs, differing for example in particle size and shape, mesopore size, and surface functionalization, as well as the animal models used, the amount administered, and the means for particle detection. The present review is an attempt to summarize the literature to date with main focus on the increasing number of studies related to quantitative full body distributions. Whenever possible, attempts are also made to discuss differences in experimental observations between studies. Finally, an outlook is given listing some open issues, and highlighting the need for more standardized experimental designs in order to allow for a faster identification of optimal particle characteristics for drug delivery applications of MSNs.

The Enzymes, Volume 43
ISSN 1874-6047
https://doi.org/10.1016/bs.enz.2018.07.007

1. INTRODUCTION

Mesoporous silica nanoparticles (MSNs) have evolved as a promising drug carrier platform, which also can be made imageable by different imaging techniques. The developments in terms of theranostic applications of MSNs have been the subject of a number of extensive recent reviews [1–4]. In comparison to other nanoparticulate drug delivery platforms, the main strengths of MSNs in these applications are high surface areas and pore volumes allowing high drug loadings to be reached, tunable mesopore dimensions and pore shapes/connectivities, easy chemical functionalization, control of particle size and shape, hydrolytic degradability under biologically relevant conditions producing biocompatible degradation products, and good overall biocompatibility. The high drug loading capacity makes it possible to use lower nanoparticle doses as compared to other nanoparticulate drug delivery platforms exhibiting lower drug loading levels, which is important from a toxicity point of view. Furthermore, the location of the drug inside the mesopore system of the MSNs may shield the drug against enzymatic degradation until released [5, 6], and the MSNs can therefore also enhance the pharmacokinetics of the drug. MSNs have been shown to allow for delivery of many different classes of bioactives, ranging from small molecule drugs, peptides, proteins, enzymes, to DNA *in vitro*. An ever increasing number of studies also demonstrate that successful drug delivery, especially to solid tumors, is also possible *in vivo*. However, as for all other nanoparticulate drug delivery platforms, only a small fraction of the nanoparticles reach their targets, and most of the particles end up in defense organs of the body rich in macrophages, like liver, spleen. A detailed discussion about the different types of cells and the biological processes involved is outside the scope of this review, and the interested reader is referred to the in-depth discussion of nanoparticle–liver interactions presented in an excellent recent review [7]. Fast accumulation of nanoparticles in the defense organs has been linked to the adsorption of serum proteins onto the particles [8, 9]. Typical approaches for decreasing the amount of protein adsorption are functionalization of the outer surface of the MSNs with hydrophilic polymers like PEG or tuning of the surface charge through covalent attachment of silanes carrying a functional end-group, typically basic $-NH_2$ or acidic $-COOH$ groups. On the other hand it has also been suggested that protein adsorption is a prerequisite for stealth properties [10], and adsorbed proteins can also enhance the cellular selectivity of actively targeted MSNs probably due to enhanced multivalency through combined specific ligand- and protein-mediated receptor interactions [11]. As outlined in more detail below, the composition of the protein corona has also been suggested to influence the biodistribution of nanoparticles. Thus, both the extent of protein adsorption and the exact composition of the protein corona are important for the *in vivo* fate of MSNs. To date, however, reports where the composition of the protein corona, the extent of protein adsorption, and the biodistribution of MSNs have been studied in parallel are scarce. Furthermore, as the experimental design of the biodistribution studies varies greatly in terms of MSN size and shape, MSN surface chemistry, administered dose, and animal model used, the current level of understanding of factors influencing the biodistribution and blood circulation times of MSNs is limited. The aim of this review is to summarize the outcome of studies aiming at an at least semi-quantitative analysis of the biodistribution of MSNs. Therefore the focus is put on studies where radiolabeled MSNs have been used, but some studies using other means of determining the biodistribution of MSNs, including MRI, ICP, and fluorescence microscopy are also discussed where appropriate. Studies using non-porous silica nanoparticles are not included with some few exceptions, as the dissolution kinetics of non-porous silica nanoparticles is clearly slower than that of MSNs of similar size, and therefore their biological fate, especially in terms of excretion kinetics [12], differs from that of MSNs.

2. RADIOLABELING OF MSNs

Radiolabeling of MSNs either with positron- or electron-emitting isotopes can be achieved using two main approaches. Either a ligand that strongly complexes the radioisotope through multiple coordinative bonds can be covalently linked to the MSNs, typically through silane chemistry, or the radioisotope can be directly linked to the silica network through interactions with deprotonated surface silanols. Either approaches have their pros and cons. The complexes used for radioisotope immobilization, typical examples being desferrioxamine B (DFO), diethylenetriaminepentaacetic acid (DTPA), and 1,4,7,10-tetraazacyclododecane-1,4,7,10-tetraacetic acid (DOTA), are already in clinical use. Efficient complexing agents are available for most clinically relevant isotopes, and limited leakage of the radioisotope can be expected, due to their multivalent coordination. For example, Meijs and co-workers reported that DFO exhibits rapid and efficient labeling with a 1:1 ratio of zirconia to chelate and demonstrated good stability with regard to demetalation, releasing less than 0.2% of the metal in serum after 24h [13]. However, the inherent need for surface modification of the MSNs, which can have an influence on the protein corona and also on the dissolution kinetics of the MSNs, may influence the biodistribution of such particles in comparison to that of the native particles. Direct introduction of the radioisotope into the silica network can be expected to have a lesser influence on these properties, although the zeta-potential and also the dissolution kinetics of the MSNs could be modified upon radiolabeling to an extent determined by the loading level, and thus the biological fate of such particles may be influenced by the introduced radiolabel. In both cases, but probably to a larger extent for directly modified particles, release of the radioisotopes from the MSNs upon MSN dissolution may be another factor that influences the results. Furthermore, the direct isotope labeling approach is most efficient for multivalent cationic isotopes, like Ti^{4+} and Zr^{4+} that have an inherent chemical compatibility with the silica network.

Typical positron emitting isotopes that have been used to label MSNs for positron emission tomography (PET) studies are ^{64}Cu and ^{89}Zr, and in one case ^{45}Ti. These isotopes all exhibit a longer half-life than the clinically often used ^{18}F, and are therefore suitable for longer term imaging of MSNs. Positron emission is attractive, as the two photons produced when a positron meets an electron are orthogonal to each other, and therefore the background signals can be efficiently reduced. This makes it possible to perform semi-quantitative imaging *in vivo*, but the most quantitative results are obtained by organ analysis post-mortem, as in the case of electron-emitting isotopes.

3. BIODISTRIBUTION OF MSNs

Quantitative or semi-quantitative studies of the biodistribution of intravenously injected MSNs have been carried out using both healthy and tumor-bearing mice. As the presence of tumors in most cases does not seem to influence the overall biodistribution, the first part of the discussion will focus on the general biodistribution of MSNs aiming at establishing structure–activity relationships wherever possible. Some of the main factors influencing the biodistribution of MSNs are summarized in Fig. 1. The second part of the discussion is focused on passive and active targeting of MSNs to tumors using intravenous administration based on studies where the tumor uptake has been studied in combination with the biodistribution

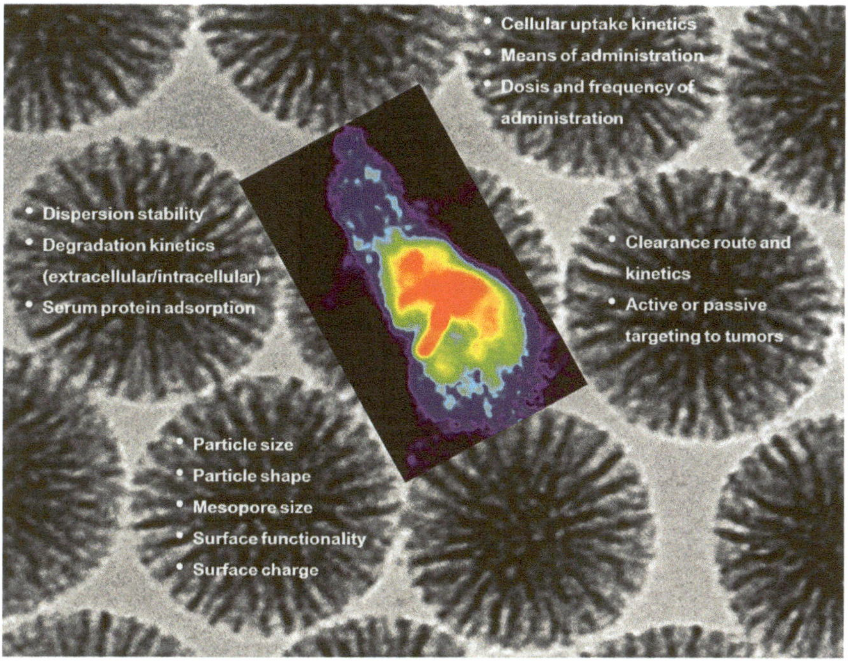

Fig. 1 Some important factors influencing the biodistribution, therapeutic efficiency, and clearance of mesoporous silica nanoparticles.

of the particles. Also here, the aim is to establish structure–activity relationships wherever possible, not only focusing on the MSNs but also on the influence of the biological model on the results. A summary of the experimental details of the studies to be discussed is given in Table 1, including mouse models, injected doses, and particle characteristics.

3.1 Overall Biodistribution of MSNs: Spherical Particles

The two studies published to date related to quantitative studies of the biodistribution of MSNs that probably are carried out under close-to-identical conditions are the studies published by Chen et al. [16] and Goel et al. [15]. Here PET-detectable, PEGylated (5k) ^{64}Cu-DOTA-tagged MSNs with a particle diameter of 80 nm have been studied using either BALB/c mice bearing xenografted 4T1 murine breast cancer tumors [16] or female athymic nude mice bearing xenografted human glioblastoma (U87MG) tumors [15] in the front flank as the animal models. Furthermore, the biodistribution of corresponding particles to which a targeting ligand had been attached, TRC105 antibody in the case of 4T1 tumors and the vascular growth factor

Table 1 Summary of the Experimental Designs Used in (Semi-)quantitative MSN Biodistribution Studies

Mean Particle Diameter (nm)	Particle Shape and Porosity	Pore Size (nm)	Functionalization	Mouse Model	Tumor Size	Particle Dose	Main Biodistribution Pattern	Means of Detection	Reference
200	Hollow, spherical particles with mesoporous shell	2–3	PEGylated (5k), Sunitinib used as tumor targeting ligand	Athymic nude mice bearing xenografted U87MG tumors in the front flank	6–8 mm	N/A, 5–10 MBq	Highest accumulation in liver (about 25%ID/g at 0.5 h, which decreases over time). Blood: about 3% ID/g at 0.5 h which remains stable up to 18 h. Higher amount of particles in tumor when actively targeted (max. 8% ID/g)	^{64}Cu-PET	[14]
80	Spherical, mesoporous	2–3	PEGylated (5k), $VEGF_{121}$ used as tumor targeting ligand	Athymic nude mice bearing xenografted U87MG tumors in the front flank	4–6 mm	N/A, 5–10 MBq	Highest accumulation in liver (about 25%ID/g at 0.5 h, which decreases over time). Blood: about 2–5% ID/g depending on if actively targeted or not at 0.5 h which remains stable up to 22 h. Higher amount of particles in tumor when actively targeted (max. 2.9% ID/g)	^{64}Cu-PET	[15]

80	Spherical, mesoporous	2–3	PEGylated (5k), TRC105 used as tumor targeting ligand	Female BALB/c mice bearing 4T1 breast cancer tumors in the front flank	5–8 mm	N/A, 5–10 MBq	Highest accumulation in liver (about 16–22%ID/g at 0.5 h, which decreases over time). Blood: about 2–5% ID/g depending on if actively targeted or not at 0.5 h which remains stable up to 22 h. Higher amount of particles in tumor when actively targeted (max. 5.9% ID/g)	^{64}Cu-PET	[16]
290	Hollow spherical particles with mesoporous shell	5–10	Native silica particles filled with perfluoro-15-crown-5 ether (PFCE)	Healthy nude MRI mice	—	200 mg/kg	Exclusive accumulation in liver 24 h p.i. No signs of remaining particles 72 h p.i.	^{19}F-MRI	[17]
76	Hollow spherical particles with mesoporous shell covered by a thin non-porous silica shell	—	PEGylated (5k) silica particles filled with perfluoro-15-crown-5 ether (PFCE)	Female BALB/cA bearing colon-26 tumors	About 2400 mm^3	N/A, 0.20 mL, $C_{PFCE} = 2.6$ mM	Exclusive accumulation in tumor and liver. No time point given	^{19}F-MRI	[30]

Continued

Table 1 Summary of the Experimental Designs Used in (Semi-)quantitative MSN Biodistribution Studies—cont'd

Mean Particle Diameter (nm)	Particle Shape and Porosity	Pore Size (nm)	Functionalization	Mouse Model	Tumor Size	Particle Dose	Main Biodistribution Pattern	Means of Detection	Reference
150	Spherical, mesoporous	8	Benzyl ferrioxamine (DFO)	Male SCID mice carrying xenografted LNCaP C4-2 tumors dorsal in the subcapsular region	Max. 480 mm^3	0.5 mg/kg	Fast removal from the blood stream. Mainly accumulating in liver and spleen 1 h p.i. Limited accumulation in tumor	^{89}Zr-PET	[18]
150	Spherical, mesoporous	8	PEGylated (5k)	Male SCID mice carrying xenografted LNCaP C4-2 tumors (left) and PC-3 tumors (right) dorsal in the subcapsular region	Max. 500 cm^3	0.5 mg/kg	Long blood circulation time (>50%ID/g 1 h p.i. and about 9%ID/g 12 h p.i.). Mainly accumulating in spleen and liver. Increasing accumulation in tumors over time (PC-3 3.8%ID/g, LNCaP 8.6%ID/g, max. time point studied 12 h)	^{89}Zr-PET	[19]
Cross-section 120–140 nm	Both spherical and elongated mesoporous silica nanoparticles (aspect ratios 1, 1.5, and 5)	2–3	Native and PEGylated (5k)	N/A	—	20 mg/kg	>80% trapped in liver, spleen, lung 2 h p.i. Elongated particles exhibited longer blood circulation times and PEGylated particles a slower clearance from the RES	ICP, fluorescence	[20]

Cross-section 120–150 nm	Both spherical and elongated mesoporous silica nanoparticles (aspect ratios 1 and 8)	3	Native and amino-functionalized particles	Healthy female CD-1 mice	—	20 mg/kg	Accumulation in lung, liver, spleen. Concentration in lung decreases over time. Higher accumulation in lung for elongated particles as compared to corresponding spherical particles, and for native as compared to amino-functionalized particles	^{125}I, radioactivity	[18]
130	Spherical, mesoporous		Amino-, phosphonate and folic acid (targeting ligand) functionalized particles	Female BALB/cAnNCrj-nu nude mice bearing xenografted human MCF-7 tumors in the left lateral abdominal	3 mm	About 50 mg/kg	More particles in lung than in liver and spleen up to 48 h p.i. Higher accumulation of particles in spleen than in liver. Actively targeted particles accumulate to a much higher degree in the tumor as compared to non-targeted particles	ICP and *in vivo* fluorescence microscopy	[21]

Continued

Table 1 Summary of the Experimental Designs Used in (Semi-)quantitative MSN Biodistribution Studies—cont'd

Mean Particle Diameter (nm)	Particle Shape and Porosity	Pore Size (nm)	Functionalization	Mouse Model	Tumor Size	Particle Dose	Main Biodistribution Pattern	Means of Detection	Reference
80–360 nm	Spherical, mesoporous particles of different sizes	N/A	Native silica and PEGylated particles (10k)	Healthy ICR mice	—	N/A	PEGylated particles exhibit a longer blood circulation time. Most particles were accumulating to a larger extent in spleen as compared to liver. Native silica particles were initially accumulating in the lung (30 min p.i.) but re-distributed with time. Stronger particle uptake by RES for larger particles	Fluorescence spectroscopy	[22]
160	Spherical mesoporous particles	12	PEGylated (5k) mesoporous silica particles to which the TRC105 antibody is covalently linked	BALB/c mice bearing xenografted 4T1 breast cancer tumors in the front flank	5–8 mm	N/A, 7.4 MBq	Comparable %ID/g values in liver and spleen, and the concentration in liver increases with time (max. 48 h). Actively targeted particles accumulate to a much higher degree in the tumor as compared to non-targeted particles	^{89}Zr-PET	[23]

50–100	Spherical	2–3	Amino-functionalized, cationic	Male nude mice and male Sprague Dawley rats	—	12–30 mg/kg	Almost exclusive accumulation in the liver short-term	NIR-fluorescence	[24]
100 nm	Spherical		PEGylated (2k) or DMPC lipid bilayer coated	Female C57BL/6 mice	—	40 mg/kg	Lipid-coated particles accumulate strongly in the liver and less in the lung as compared to native or PEGylated particles	ICP-MS	[25]
Spherical particles 200 nm, rod-shaped: cross-section 120–150 nm	Both spherical and elongated mesoporous silica nanoparticles (aspect ratios 1, 2 and 4)	3	Magnetic mesoporous silica particles. No information about surface functionalization	BALB/c nu/nu mice bearing HepG2 tumors on one side of the dorsal flank	N/A	5 mg/kg	Elongated particles accumulated more efficiently in the tumor as compared to spherical particles	ICP-OES, CLSM	[26]
150 nm	Spherical	3	PEGylated (5k)	BALB/c mice	5–8 mm	11.1 MBq	Long blood circulation time, about 5%ID/g 20 h p.i.	^{45}Ti-PET	[27]

$VEGF_{121}$ in the case of U87MG tumors, was also studied. The MSN dose is not reported, but in both studies the injected dose was 5–10 MBq, suggesting comparable doses in both studies. The distribution of the MSNs in liver, tumor, blood and muscle was followed over time, and a full organ distribution was determined at the end of the experiment (22 h post-injection (p.i.) for mice carrying U87MG tumors and 48 h p.i. for mice carrying 4T1 tumors). The observed distributions were similar in both cases and the presence of a targeting ligand had only minor influences on the general biodistribution. The highest concentrations of particles were observed in the liver, 16–24%ID/g 0.5 h p.i., which gradually decreased over time to reach 12–16%ID/g 22–24 h p.i. Minimal accumulation was observed in muscle, <1%ID/g. In both studies the particle concentration in the blood remained relatively low and constant during the time of observation, being about 2–4%ID/g. The full organ analysis post-mortem also showed the highest particle content in the liver, and lower particle concentrations in lung, spleen, kidney, and intestines. Thus, the biodistribution was almost identical despite that different mouse models were used. The particle concentrations in the blood and in the muscles remained unaffected by the introduction of a targeting ligand. However, in the case of TRC105-tagged MSNs, an increased accumulation in the liver was observed as compared to that of the non-tagged particles, while this was not observed for the $VEGF_{121}$-tagged particles. This observation may reflect targeting ligand-dependent changes in the protein corona of the particles. However, as no protein corona analysis was carried out, this suggestion remains speculative at this stage. In a related study using 160 nm MSNs intrinsically ^{89}Zr-labeled, PEGylated (5k) MSNs equipped with TRC-105 antibodies and using the same 4T1 animal model as discussed above, Goel et al. again observed a pronounced uptake in liver and spleen as analyzed 48 h p.i. [23]. However, the uptake in the spleen was now at the same level as that of the liver, which stands in contrast to the results obtained for the corresponding 80 nm particles. Similar results were reported by Chen et al. for corresponding PEGylated (5k) and intrinsically ^{45}Ti-labeled, PET active MSNs but without targeting ligand [27]. Thus, it appears that increasing the particle size from 80 to 160 nm for PEGylated (5k) MSNs leads to an increase in the relative number of particles taken up by the spleen as compared to the liver.

The biodistribution of 80 nm spherical native and PEGylated (10k) MSNs has also been studied by He et al. using tumor-free ICR mice as the model [22]. No information about the administered amount of particles was given, and fluorescence spectroscopy was used for biodistribution analysis using homogenized tissues. In contrast to the results reported for the 80 nm MSNs discussed above, the highest particle concentrations were observed in the spleen for both type of particles 30 min p.i. (μg MSNs/mg), but particles were also present in liver and lung. The initial concentrations of MSNs in the blood were high, but decreased exponentially over time to low values 8 h p.i. The organ concentrations of the PEGylated particles were lower than that observed for the non-PEGylated particles throughout, which may suggest that PEGylation prolongs the blood circulation time of the MSNs. The particle concentration in these organs, and especially so in the lungs, decreased over time, which can be seen as combined effect of particle excretion and re-distribution over time. In the same study the effect of particle size was also studied for 80, 120, 200, and 360 nm native and PEGylated particles. As expected, PEGylation generally enhanced the blood circulation time regardless of the particle size. However, the blood circulation time decreased with increasing particle size, which was mirrored by an increase of the particle concentrations in liver and spleen for the non-PEGylated particles. Interestingly, a similar particle size dependency, albeit with lower absolute concentrations values, was observed for the accumulation of PEGylated particles in the spleen during the first days p.i. for particle sizes in the range 80–200 nm, while a clearly lower particle concentration in the spleen was observed for the 360 nm particles. The exact reason for this observation remains unclear, as no serum protein adsorption results were provided, but it is fair to assume that the lower uptake in the spleen for the largest particles studied is connected to the much shorter blood circulation time observed for these particles as compared to the smaller particles.

The effect of PEGylation (5k) was also studied by Kramer et al. for PET imageable ^{89}Zr-DFO-MSNs having a mean particle diameter of 150 nm [19]. Male SCID mice carrying both dorsally xenografted LNCaP (C4-2, left side) and PC-3 tumors (right side) were used as the animal model. As the particles were very bright, the injected dose could be kept very low, 0.5 mg/kg. The non-PEGylated particles accumulated very fast in lung, liver, and spleen (<1 h p.i.) [19, 28]. The amount of particles accumulating in all three organs increased over time until 3 h p.i., the latest time point studied for these particles, reaching 28%ID/g (lung), 64%ID/g (liver), and 53%ID/g (spleen). The blood concentration of the particles remained low throughout, being below 1%ID/g. PEGylation of these particles led to a drastically increased blood circulation time, reaching values as high as 32%ID/g 1 h p.i., and 6%ID/g 12 h p.i. Interestingly, a pronounced particle accumulation in the spleen was observed, 130%ID/g 1 h p.i. which further increased to 280%ID/g 12 h p.i. These values were much higher than those observed for the liver (12%ID/g 1 h p.i. and 15%ID/g 12 h p.i.). Thus, the higher organ mass normalized uptake in the spleen as compared to liver is in good agreement with the results of He et al. [22], but stands in contrast to the observations by Chen et al. [27] and Goel et al. [15]. The differences are most probably reflecting the differences in the blood circulation times of the particles, where a longer blood residence time of the particles leads to a higher uptake in the spleen. An additional reason for such an observation could be that the amount of particles that the spleen can process at the same time is limited, and the very low particle loading used in the study of Kramer et al. [19] can therefore also lead to a higher relative number of particles in the spleen. However, a very high relative uptake of PEGylated particles by the spleen has also been observed by Nissinen et al. for related mesoporous silicon nanoparticles [29], which in their study was attributed to a lower amount of serum protein adsorption to dually PEGylated particles in combination with a changed relative composition of the protein corona upon PEGylation. No full serum protein adsorption results were reported by Kramer et al. [19], but a drastic drop in the adsorption of BSA upon PEGylation was demonstrated. The particle content in the lung was about 10%ID/g 1 h p.i. which decreased gradually to about 2%ID/g 12 h p.i., suggesting that the particles re-distributed over time. Furthermore, while the particle concentration in the liver, lung, and spleen decreased over time in the study of He et al. [22], an increase in the particle concentration in liver and spleen in combination with a decreasing concentration of particles in the lung was observed by Kramer et al. [19]. Thus, one can suggest that also this observation can be connected to the differences in blood circulation time in the two studies. Kramer et al. [19] further related the particle concentrations in the organs to the concentration of particles in the blood. A linear correlation passing origo was observed for heart, brain, lung, and muscle, suggesting that no particle accumulation occurred in these organs, as shown in Fig. 2. For liver, spleen and kidney a negative correlation was observed, in agreement with particle accumulation in these organs over time. These results also highlight the need for taking particles present in the blood still remaining in the organs into account before conclusion about particle accumulation in organs can be made.

In an interesting study using ^{125}I radiolabeled 100 nm MSNs either in native or amino-functionalized form, Yu et al. studied the biodistribution using immune-competent CD-1 mice [18]. Up to 2 h p.i. the highest

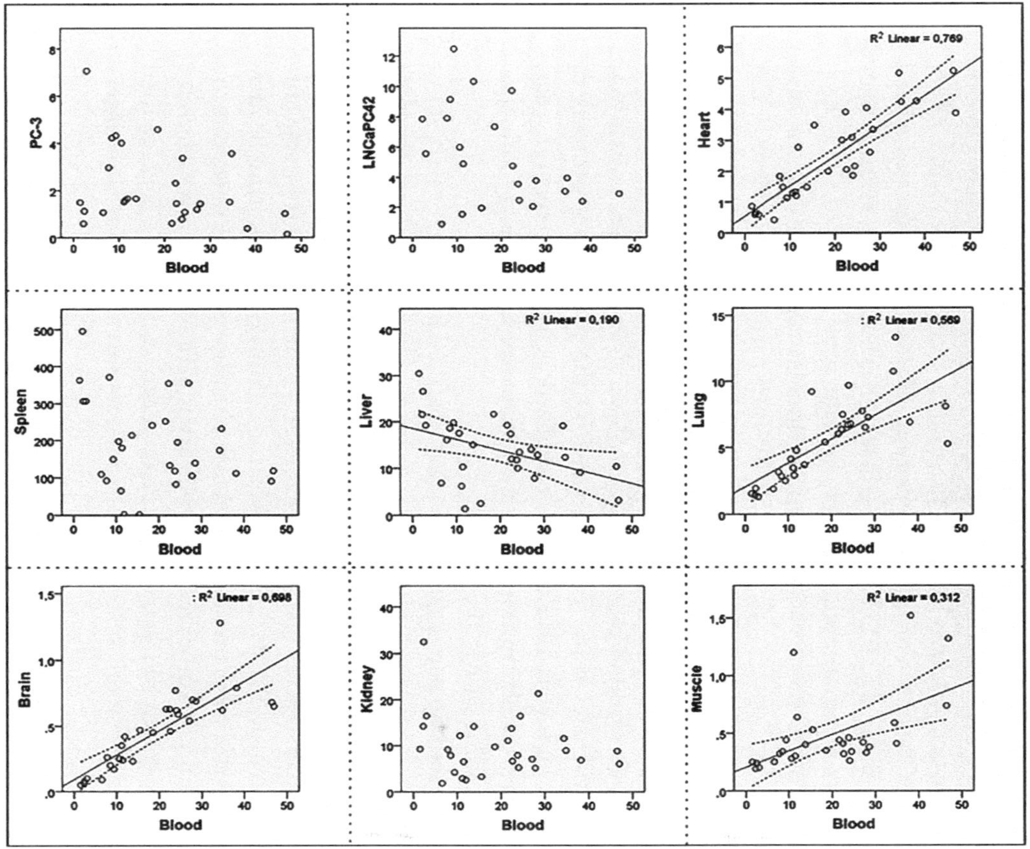

Fig. 2 Correlation between blood concentration and organ concentrations of MSNs (diameter 150 nm) within the first 12 h after intravenous injection into the tail vein of mice at a dose of 0.5 mg/kg. See Kramer et al. for details. *Reproduced by permission from the Royal Society of Chemistry, L. Kramer, G. Winter, B. Baur, A.J. Kuntz, T. Kull, C. Solbach, A.J. Beer, M. Lindén, Quantitative and correlative biodistribution analysis of 89 Zr-labeled mesoporous silica nanoparticles intravenously injected into tumor-bearing mice, Nanoscale 27 (2017) 9743–9753.*

particle concentration was observed in the lung for the native particles, followed by spleen and liver. For corresponding (cationic) amino-functionalized particles the concentration of particles in the lung was minimal, while a very pronounced accumulation of the particles in the spleen (150%ID/g) followed by liver (about 28%ID/g) was observed. The rather different overall biodistribution pattern observed in this study as compared to those observed in the studies discussed above may be connected to differences in particle loadings between the studies, but also to probable differences in the protein corona composition of the amino-functionalized particles as compared to PEGylated particles. However, this remains a speculation as no serum protein adsorption data were provided. The biodistribution of strongly cationic, amino-functionalized MSNs has also been studied by Souris et al. using 50–100 nm particles [24]. Here, a fast and virtually exclusive particle accumulation in the liver was observed, which was ascribed to enhanced protein adsorption onto the strongly cationic MSNs. A fast and high liver accumulation has also been observed by Rascol et al. for lipid bilayer-coated MSNs with a diameter of 100 nm [25]. These particles were less influenced by the presence of proteins in the medium and were also faster internalized *in vitro* by Hep-G2 cells, suggesting that cellular internalization kinetics may also strongly influence the biodistribution of MSNs.

A biodistribution pattern similar to that observed by Souris et al. [24] but with slower accumulation kinetics has been observed in two independent studies for hollow MSNs filled with PFCE, a perfluorinated ^{19}F-MRI contrast agent which is a liquid that is virtually insoluble in water [17, 30]. Here, different particle sizes (mean particle diameter: 76 nm *versus* 290 nm) and animal models (female BALB/cA mice bearing colon-26 tumors *versus* healthy nude MRI mice) were used. In both cases virtually exclusive uptake in the liver (and tumor) was observed. This very atypical biodistribution pattern for anionic, native silica MSNs can only be understood if the PFCE is assumed to play an active role in influencing the interactions at the bio-MSN interface. In the study by Pochert et al. [17] the hollow MSNs had a mesoporous shell with open mesopores allowing direct contact between PFCE and the surrounding. Analysis of the protein adsorption from full serum revealed that the PFCE-filled particles adsorbed clearly higher amounts of apolipoproteins A-1 and A-2 as compared to the same particles not filled with PFCE. A-1 is known to guide the transport of cholesterol to the liver [31], and was recently exploited for targeting liposomes to the liver [32]. Therefore it was suggested that the unusual biodistribution observed for these particles was connected to PFCE-induced changes in the protein corona composition of the particles. In the study of Nakamura et al. [30] on the other hand, a 10 nm thin coating of non-porous silica was deposited on the particles after PFCE loading. Thus, a biodistribution pattern similar to that of native silica nanoparticles would be expected, with pronounced uptake also in the spleen. However, as amorphous silica dissolves relatively rapidly under physiological conditions, such a thin silica layer can dissolve and expose PFCE to the biological system leading to the above discussed influences on the protein corona of the particles. However, an additional influence of amount of particles administrated on the biodistribution cannot be excluded, as very high particle concentrations were applied in both studies due to contrast issues.

3.2 Overall Biodistribution of MSNs: Influence of Particle Shape

The influence of particle shape on the biodistribution of MSNs has also been studied. Huang et al. compared the biodistribution of non-PEGylated and PEGylated (5k) [20]spherical, and elongated (aspect ratio (AR) 1.5 and 5) MSNs with a cross-section of about 140 nm. The biodistribution analysis was based on a combination of fluorescence microscopy and ICP-OES analyses, but the exact mouse model was not specified. Yu et al. compared ^{125}I radiolabeled native and amino-functionalized spherical and elongated (AR 8) MSNs with a cross-section of 100 nm using an immunocompetent CD-1 mouse model [18]. In both cases a particle loading of 20 mg/g was used. Although the results are not fully conclusive, the elongated native silica particles seem to accumulate to a larger extent in the spleen as compared to their spherical counterparts, at least short term (a few days), but cylindrical particles seem to have a shorter blood circulation time as compared to spherical particles exhibiting a similar cross-section. Similar conclusions were also reported by Shao et al. for mesoporous silica nanoparticles about similarly sized but with a lower dose of 5 mg/kg [26]. Amino-functionalization leads to a pronounced initial entrapment in the lungs as compared to that observed for corresponding spherical particles, but these particles re-distributed fast and were entrapped more in the spleen than in the liver (mass/mass). Thus, it appears that the entrapment of particles in the lung is at least partly of physical nature, and the extent of lung entrapment is a function both of particle functionalization and particle shape.

3.3 Passive *Versus* Active Targeting of MSNs in Connection to Circulation Time

A bulk of studies have focused on either passive or active (receptor-mediated) targeting of MSNs to tumors *in vivo*. For a more detailed discussion of the therapeutic aspects of these studies, consultation of recent reviews is suggested. The aim of the following discussion is to summarize the literature related to tumor uptake, and to put the tumor uptake in relation to the overall biodistribution of the same particles.

In the studies of Chen et al. [16] and Goel et al. [15] using PEGylated 80 nm spherical MSNs discussed above, passive *versus* active targeting was analyzed. The MSNs were then equipped with ligands actively targeting the cancer cells, TRC105 antibodies in the case of 4T1 tumors and the vascular endothelial growth factor $VEGF_{121}$ in the case of U87MG tumors. Clear differences were observed in the tumor accumulation pattern in the two cases. A fast accumulation of the TRC105-tagged particles in the 4T1 tumors was observed, reaching about 6%ID/g 0.5 h p.i. and which stayed constant until 5 h p.i. after which it gradually decreased. For corresponding particles not carrying a targeting ligand a relatively time-independent particle concentration in the tumor of 2.5–3%ID/g was observed. A similar passive tumor uptake of the PEGylated MSNs was also observed for the U87MG model, in agreement with the similar and constant particle concentration in the blood over time in both cases, 2–3%ID/g. However, here a clearly higher influence of active targeting was observed as compared to the 4T1 model, but with similar particle accumulation kinetics. A fast particle accumulation was observed which peaked at about 8%ID/g 6 h p.i., after which a gradual decrease in the particle content of the tumor was observed. Close to identical results were also reported for hollow 150–250 nm MSNs carrying a peptide-based targeting ligand, cRGDyK [14], suggesting that the observed differences between the active targeting efficiencies for the 4T1 model as compared to the U87MG model may be model dependent rather than particle and targeting ligand dependent, as will be discussed in more detail below. Increasing the particle size to 160 nm for the TRC105-tagged particles led to clear changes in the 4T1 tumor accumulation pattern. Here, the tumor accumulation was slower, but reached as high tumor accumulation as 11.4%ID/g 6 h p.i., after which it remained relatively constant up to 48 h p.i. However, also the accumulation of the corresponding particles not carrying the TRC105 targeting ligand was slower than that observed for the 80 nm particles and peaked 24 h p.i. at about 3.5%ID/g, which is a similar value as that observed for the passively targeted 80 nm particles. Keeping in mind that the larger particles evidently in this case showed a longer blood circulation time, as also suggested by the higher spleen/liver ratio of the 160 nm particles as compared to the 80 nm particles, a more efficient active targeting over time can be achieved. In an additional study, the uptake of passively targeted, PEGylated (5k) 160 nm MSNs by 4T1 tumors was studied, and also here a slower tumor accumulation was observed as compared to comparable 80 nm particles, reaching 4.6%ID/g 6 h p.i., in line with the above conclusions [27].

Passive accumulation of MSNs in two different prostate cancer tumors present in the same mice was studied by Kramer et al. [19]. Here PEGylated (5k) 160 nm MSNs that exhibited a very long blood circulation time were used. The uptake in both type of tumors increased continuously with time during the whole period of observation (end point; 12 h p.i.), reflecting the value of long blood circulation times for efficient passive particle uptake in the tumor. The MSN uptake in the tumors increased with time to reach values as high as 8.6%ID/g in the LNCaP tumors, while the corresponding value for the PC-3 tumors was 3.8%ID/g 12 h p.i. An increase in particle concentration in solid 4T1 murine breast tumors with time for passively targeted PEGylated MSNs has also been reported by Chen et al. [27]. Based on *in vitro* cell culture experiments the opposite uptake behavior would have been expected. This observation could be connected to the difference in vascularization of the two tumors, as the LNCaP tumors were clearly more vascularized than the PC-3 tumors. Thus, this study clearly shows that a direct comparison of absolute %ID/g values between studies is of limited value when it comes to estimating the general efficiency of any given particle type for tumor targeting, unless the exact same experimental conditions have been applied.

Especially high tumor accumulation was reported in the study of Lu et al. using phosphonated, folate-targeted particles to actively target human, xenografted MCF-7 breast cancer tumors [21]. Folate targeting led to enhanced tumor accumulation of MSNs even 48 h p.i. Highly efficient folate targeting to MDA-MB-231 breast cancer tumors has also been reported [5], suggesting that folate targeting breast cancers exhibiting high concentrations of folic acid receptors is very efficient. Interestingly, a higher MSN accumulation in spleen than in liver was seen in the study by Lu et al. both for the native MSNs and the particles carrying folate [21], and a high tumor uptake was also observed for the folate-free particles albeit not being as high as for the folate-tagged MSNs. This suggests that the high tumor uptake is related to a relatively long blood circulation time of these particles, a prerequisite for an efficient active targeting over time.

Taken together, these results stand in stark contrast to the results presented by Goel et al. and Chen et al., and most other comparable studies, where a fast particle accumulation mainly in the liver is observed. As already discussed above, these differences can be attributed to the large differences in the blood circulation times observed in the different studies, and also suggest that "PEGylation" as such is not the sole important factor for biodistribution, but more importantly the efficiency by which the PEGylation influences protein adsorption onto the MSNs, which apart from the number and identity of proteins adsorbed could also influence the conformation of the proteins (or some of the proteins), and thus their bioresponse [33]. Unfortunately, such results are seldom reported in connection to the biodistribution studies, which makes a more detailed discussion of the influence of this parameter impossible at this stage.

4. EXCRETION OF MSNs

The degradability and clearance of silicon-based nanoparticles has recently been extensively reviewed [34], which is why only a brief summary focused specifically on MSNs is given here. Renal clearance has identified as the main excretion route for most types of MSNs, although particles are also excreted hepatobiliary but to a lower extent and typically with a slower kinetics [20]. For example, using phosphonate-functionalized, spherical MSNs 100–130 nm in diameter Lu et al. [21] showed that most MSNs were observed in spleen, primarily in the red pulp, where red cells, phagocytes, and macrophages are abundant, and in the liver mainly in the portal vein 4 h p.i., before the vast majority of the particles were renally excreted and a minority excreted through feces within 96 h. The renal clearance cutoff for globular filtration is about 6 nm [35, 36], which is why partial particle dissolution can be assumed to be a prerequisite for this means of clearance. On the other hand, if the particles are transported through the epithelial cells lining the renal tubules into the tubular fluid, a cutoff size of 70–90 nm has been suggested [12]. As most MSNs studied *in vivo* to date have diameters exceeding 70–90 nm, a partial particle dissolution must be assumed to be a prerequisite also in this case. Surprisingly, however, intact MSNs having particle dimensions exceeding 90 nm have been observed in the urine [20, 21, 37], a result for which there is currently no widely explanation. Hepatobiliary excretion goes through the liver and bile, and it is known that protein adsorption plays a key role in directing nanoparticles to the liver [38]. Furthermore, the reticuloendothelial system can phagocytize particles and then transport them to the liver, before being translocated to the bile and then further excreted in the feces. Aggregation of small particles has also been shown to direct the excretion toward excretion through feces, highlighting the influence of the effective particle dimensions on the preferred route of excretion [39].

Elongated MSNs also appear to, at least initially, be mainly renally cleared, accompanied by hepatobiliary clearance with slower kinetics [22]. However, as compared to corresponding spherical particles, the clearance rate of the elongated particles was slower.

However, there seems to be a connection between the blood circulation time of the MSNs and the excretion route. Using 50–100 nm, cationic MSNs exhibiting a very short blood life time due to a fast particle accumulation in the liver, Souris et al. demonstrated a fast and virtually exclusive hepatobiliary clearance of the particles, which was connected to the high positive charge of the amino-functionalized MSNs used in this study [24]. This result stands in contrast to most other studies typically using PEGylated MSNs exhibiting a longer blood circulation time, where renal clearance is initiated virtually immediately after administration and typically peaks 1–2 days p.i., after which the concentration of particle in the feces reaches a maximum (single dose administration). The clearance rate of PEGylated particles has been shown to be slower than that of corresponding native particles [22], which in addition to PEGylation related enhanced stealth properties can be related to the slower dissolution rate of PEGylated particles [40]. For more slowly dissolving non-porous silica nanoparticles having sizes clearly exceeding those needed for an efficient globular filtration, long clearance times on the time-scale of months have been observed [12], highlighting the immense importance of silica nanoparticle degradation kinetics, extra- and intracellular, for determining the excretion kinetics and probably also clearance route of the particles *in vivo*.

5. CONCLUSIONS AND OUTLOOK

To date, the knowledge about the factors influencing the biodistribution of MSNs is relatively shattered due to the relatively limited numbers of studies where serum protein adsorption, particle dissolution in serum, and (semi)-quantitative biodistribution experiments have been performed using the same particles. Furthermore, the variation in experimental design between the different biodistribution studies is broad (different animal models, differences in particle size and shape, different surface functionalization, varying particle dissolution rates, different protein corona compositions, and different administered doses), which makes it still challenging to define particle design criteria for achieving an optimal biodistribution of MSNs for a given application, despite an ever increasing number of *in vivo* studies published. Furthermore, often the biodistribution is not the main focus of these *in vivo* studies, but rather to show an enhanced anti-cancer activity of drug-loaded MSNs in comparison to the free drug, and also here information about protein adsorption, dissolution kinetics in serum, *etc.*, is missing. In addition, the kinetic nature of the particle mechanical stability, which naturally is related to MSN dissolution kinetics, the time-dependent evolution of the composition of the protein corona, the influence of flow, possible changes in particle polarity and charge over time due to incongruent particle dissolution including detachment of originally covalently attached functional moieties, *etc.*, makes it especially challenging to connect the physicochemical characteristics of the initially administrated MSNs to their long-term (hours to days) biodistribution. Another complication arises from the large particle-to-particle variations in terms of protein corona composition, as recently shown for MSNs using super-resolution microscopy [41], and any "mean" value for protein corona composition is bound to fail to fully describe the biodistribution for any set of MSNs. However, despite these shortcomings, many of which will be very difficult to experimentally study, some general trends and implications can be established.

It is well-established that prolonged circulation times can be achieved by proper functionalization of the outer surface of nanoparticles, typically using hydrophilic polymers like PEG and by particle size control. Also particle shape can be used to influence biodistribution and clearance kinetics, but based on the limited number of studies published to date using spherical and elongated particles, a clear overall benefit from using elongated particles as compared to spherical particles is not immediately evident. A longer circulation time of MSNs appears to lead to an increase of the spleen-to-liver ratio, as could be expected due to the size differences of these two organs. However, it is clear from the discussion above that PEGylated MSNs can exhibit clearly different circulation times although the particle sizes are comparable. It has been shown that protein adsorption to a PEGylated surface will be dependent on the surface concentration and conformation of the PEG-chains [42], and even so that the conformation of an individual protein present in the corona could be influenced by the packing density of the PEG-chains [33]. Thus, one may assume that hydrolytically more stable MSNs with an optimal surface functionalization and particle size typically in the range 100–150 nm should exhibit longer blood residence times than corresponding particles that biodegrade with a faster kinetics. A long blood circulation time should also enhance the outcome of active targeting strategies and cell-directed drug delivery using such particles. On the other hand, efficient therapeutic outcomes using long-circulating MSNs would then also require that the drug is retained inside the MSNs until these are internalized by the target cells, a process which may be relatively slow for particles with a long blood circulation time. Furthermore, a slower particle dissolution/disintegration kinetics could also slow down the clearance kinetics, which could lead to negative side-effects and also put a limit to the dosing frequency possible. An alternative approach toward maximizing the drug concentration in the target cells could therefore be to use MSNs that exhibit a fast cellular uptake kinetics by the target cells as drug carriers, although the blood circulation time may be shorter for such particles. In that case, MSNs with a faster degradation kinetics, and therefore also clearance kinetics, would be preferable. Particles that would be stable in serum but which would degrade fast after being internalized by cells, like the mesoporous silica nanoparticles containing redox cleavable S–S-bonds as part of their network [43–45], would be especially interesting from this perspective, as this could allow a combination of long blood circulation times and a fast particle clearance kinetics from the defense organs.

The establishment of solid MSN design criteria for achieving an optimum biodistribution, clearance, and therapeutic profile is yet to be established even for relatively simple small-animal models carrying solid tumors. This is reflecting the general problems of the nanomedicine field, where the years of high promise has failed to transform into broad-band drug formulations in clinical use, and where much of the current research efforts and funding are directed toward showing enhanced therapeutic outcome in small-animal models using nanoparticulate carriers, but with individual experimental designs with regards to targets, type of particles, surface functionalization, doses, *etc.*, often using just one type of MSNs in the studies. Furthermore, often limited characterization of the particles under biologically relevant conditions (particle dissolution kinetics in serum, kinetic analysis of protein adsorption from full serum, *etc.*) is performed, making the establishment of any true structure–activity relationships difficult if not even impossible in most cases. As a solid increase in the funding for basic research along these lines will probably not broadly materialize, an increased communication and collaboration between research groups in this field is urgently needed in order to enhance the knowledge outcome from lengthy and expensive *in vivo* trials.

REFERENCES

[1] V. Mamaeva, C. Sahlgren, M. Lindén, Mesoporous silica nanoparticles in medicine—recent advances, Adv. Drug Deliv. Rev. 65 (2013) 689–702.
[2] Z. Li, J.C. Barnes, A. Bosoy, J.F. Stoddart, J.I. Zink, Mesoporous silica nanoparticles in biomedical applications, Chemistry 41 (2012) 2590–2605.
[3] L. Pasqua, A. Leggio, D. Sisci, S. Andò, C. Morelli, Mesoporous silica nanoparticles in cancer therapy: relevance of the targeting function, Mini Rev. Med. Chem. 16 (2016) 743–753.
[4] R.R. Castillo, M. Colilla, M. Vallet-Regi, Advances in mesoporous silica-based nanocarriers for co-delivery and combination therapy against cancer, Expert Opin. Drug Deliv. 14 (2017) 229–243.
[5] V. Mamaeva, J.M. Rosenholm, L.T. Bate-Eya, L. Bergman, E. Peuhu, A. Duchanoy, L.E. Fortelius, S. Landor, D.M. Toivola, M. Lindén, C. Sahlgren, Mesoporous silica nanoparticles as drug delivery systems for targeted inhibition of Notch signaling in cancer, Mol. Ther. 19 (2011) 1538–1546.
[6] K. Braun, A. Pochert, M. Lindén, M. Davoudi, A. Schmidtchen, R. Nordström, M. Malmsten, Membrane interactions of mesoporous silica nanoparticles as carriers of antimicrobial peptides, J. Colloid Interface Sci. 475 (2016) 161–170.
[7] Y.-N. Zhang, W. Poon, A.J. Tavares, I.D. McGilvray, W.C.W. Chan, Nanoparticle-liver interactions: cellular uptake and hepatobiliary elimination, Biomaterials 240 (2016) 332–348.
[8] C. Gunawan, M. Lim, C.P. Marquis, R. Amal, Nanoparticle-protein corona composition govern the biological fates and functions of nanoparticles, J. Mater. Chem. B 2 (2014) 2060–2083.
[9] F. Alexis, E. Pridgen, L.K. Molnar, O.C. Farokhzad, Factors affecting the clearance and biodistribution of polymeric nanoparticles, Mol. Pharm. 5 (2008) 505–515.
[10] F. Schottler, G. Becker, S. Winzen, T. Steinbach, K. Mohr, L. Landfester, F.R. Wurm, Protein adsorption is required for stealth effect of poly(ethylene glycol)- and poly(phosphoester)-coated nanocarriers, Nat. Nanotechnol. 11 (2016) 372–377.
[11] M. Beck, T. Mandal, C. Buske, M. Lindén, Serum protein adsorption enhances active leukemia stem cell targeting of mesoporous silica nanoparticles, ACS Appl. Mater. Interfaces 9 (2017) 18566–18574.
[12] M.A. Malfatti, H.A. Palko, E.A.K.W. Turteltaub, Determining the pharmacokinetics and long-term biodistribution of SiO_2 nanoparticles in vivo using accelerator mass spectrometry, Nano Lett. 12 (2012) 5532–5538.
[13] W.E. Meijs, J.D.M. Herscheid, H.J. Haisma, H.M. Pinedo, Evaluation of desferal as a bifunctional chelating agent for labeling antibodies with Zr-89, Appl. Radiat. Isot. 43 (1992) 1443–1447.
[14] R. Chakravarty, S. Goel, H. Hong, F. Chen, H.F. Valdovinos, R. Hernandez, T.E. Barnhart, W. Cai, Functionalized hollow mesoporous silica nanoparticles for tumor vasculature targeting and PET image-guided drug delivery, Nanomedicine 91 (2015) 1233–1246.
[15] S. Goel, F. Chen, H. Hong, H.F. Valdovinos, R. Hernandez, S. Shi, T.E. Barnhart, W. Cai, $VEGF_{121}$-conjugated mesoporous silica nanoparticle: a tumor targeted drug delivery system, ACS Appl. Mater. Interfaces 6 (2014) 21677–21685.
[16] F. Chen, H. Hong, Y. Zhang, H.F. Valdovinos, S. Shi, G.S. Kwon, C.P. Theuer, T.E. Barnhart, W. Cai, *In vivo* tumor targeting and image-guided drug delivery with antibody-conjugated, radiolabeled mesoporous silica nanoparticles, ACS Nano 7 (2013) 9027–9039.
[17] A. Pochert, I. Vernikouskaya, F. Pascher, V. Rasche, M. Lindén, Cargo-influences on the biodistribution of hollow mesoporous silica nanoparticles as studied by quantitative 19F-magnetic resonance imaging, J. Colloid Interface Sci. 488 (2017) 1–9.

[18] T. Yu, D. Hubbard, A. Ray, H. Ghandehari, In vivo biodistribution and pharmacokinetics of silica nanoparticles as a function of geometry, porosity and surface characteristics, J. Control. Release 163 (2012) 46–54.
[19] L. Kramer, G. Winter, B. Baur, A.J. Kuntz, T. Kull, C. Solbach, A.J. Beer, M. Lindén, Quantitative and correlative biodistribution analysis of 89 Zr-labeled mesoporous silica nanoparticles intravenously injected into tumor-bearing mice, Nanoscale 27 (2017) 9743–9753.
[20] X. Huang, L. Li, T. Liu, N. Hao, H. Liu, D. Chen, F. Tang, The shape effect of mesoporous silica nanoparticles on biodistribution, clearance, and biocompatibility in vivo, ACS Nano 5 (2011) 5390–5399.
[21] J. Lu, M. Liong, Z. Li, J.I. Zink, F. Tamanoi, Biocompatibility, biodistribution, and drug-delivery efficiency of mesoporous silica nanoparticles for cancer therapy in animals, Small 6 (2010) 1794–1805.
[22] Q. He, Z. Zhang, F. Gao, Y. Li, J. Shi, In vivo distribution and urinary excretion of mesoporous silica nanoparticles: effects of particle size and PEGylation, Small 2 (2011) 271–280.
[23] S. Goel, F. Chen, S. Luan, H.F. Valdovinos, S. Shi, S.A. Graves, F. Ai, T.E. Barnhart, C.P. Theuer, W. Cai, Engineering intrinsically zirconium-89 radiolabeled self-destructing mesoporous silica nanostructures for in vivo biodistribution and tumor targeting studies, Adv. Sci. 3 (2016) 1600122.
[24] J.S. Souris, C.-H. Lee, S.-H. Cheng, C.-T. Chen, C.-S. Yang, J.A. Ho, C.-Y. Mou, L.-W. Lo, Surface charge-mediated rapid hepatobiliary excretion of mesoporous silica nanoparticles, Biomaterials 31 (2010) 5564–5574.
[25] E. Rascol, M. Daurat, A. Da Silva, M. Maynadier, C. Dorandeu, C. Charnay, M. Garcia, J. Lai-Kee-Him, P. Bron, M. Auffan, W. Liu, B. Angeletti, J.-M. Devoiselle, Y. Guari, M. Gary-Bobo, J. Chopineau, Biological fate of Fe_3O_4 core-shell mesoporous silica nanoparticles depending on particle surface chemistry, Nanomaterials (Basel) 7 (2017) 162.
[26] D. Shao, M.-M. Lu, Y.-W. Zhao, F. Zhang, Y.-F. Tan, X. Zheng, Y. Pan, X.-A. Xiao, W.-F. Dong, J. Li, L. Chen, The shape effect of magnetic mesoporous silica nanoparticles on endocytosis, biocompatibility and biodistribution, Acta Biomater. 49 (2017) 531–540.
[27] F. Chen, H.F. Valdovinos, R. Henandez, S. Goel, T.E. Barnhart, W. Cai, Intrinsic labeling of titanium-45 using mesoporous silica nanoparticles, Acta Pharmacol. Sin. 38 (2017) 907–913.
[28] L. Miller, G. Winter, B. Baur, B. Witulla, C. Solbach, S. Reske, M. Lindén, Synthesis, characterization, and biodistribution of multiple 89 Zr-labeled pore-expanded mesoporous silica nanoparticles for PET, Nanoscale 6 (2014) 4928–4935.
[29] T. Nissinen, S. Näkki, H. Laakso, D. Kučiauskas, A. Kaupinis, M.I. Kettunen, T. Liimatainen, M. Hyvönen, M. Valius, O. Gröhn, V.-P. Lehto, Tailored dual PEGylation of inorganic porous nanocarriers for extremely long blood circulation in vivo, ACS Appl. Mater. Interfaces 8 (2016) 32723–32731.
[30] H. Matsushita, S. Mizukami, F. Sugihara, Y. Nakanishi, Y. Yoshioka, K. Kikuchi, Multifunctional core–shell silica nanoparticles for highly sensitive ^{19}F magnetic resonance imaging, Angew. Chem. Int. Ed. 126 (2018) 1026–1029.
[31] B. Lou, X.-L. Liao, M.-P. Wu, P.-F. Cheng, C.-Y. Yin, Z. Fei, High-density lipoprotein as a potential carrier for delivery of a lipophilic antitumoral drug into hepatoma cells, World J. Gastroenterol. 11 (2005) 954–959.
[32] S.I. Kim, D. Shin, T.H. Choi, J.C. Lee, G.-J. Cheon, K.-Y. Kim, M. Park, M. Kim, Systemic and specific delivery of small interfering RNAs to the liver mediated by apolipoprotein A-I, Mol. Ther. J. Am. Soc. Gene Ther. 15 (2007) 1145–1152.

[33] C. Bernhard, S.J. Roeters, J. Franz, T. Weidner, M. Bonn, G. Gonella, Repelling and ordering: the influence of poly(ethylene glycol) on protein adsorption, Phys. Chem. Chem. Phys. 19 (2017) 28182–28188.
[34] J.G. Croissant, Y. Fatieiev, N.M. Khasab, Degradability and clearance of silicon, organosilica, silsesquioxane, silica mixed oxide, and mesoporous silica nanoparticles, Adv. Mater. 29 (2017) 1604634–1604685.
[35] M. Yu, J. Zheng, Clearance pathways and tumor targeting of imaging nanoparticles, ACS Nano 9 (2015) 6655–6674.
[36] H.S. Choi, W. Liu, P. Misra, E. Tanaka, J.P. Zimmer, B.P. Ipe, M.G. Bawendi, J.V. Frangioni, Renal clearance of quantum dots, Nat. Biotechnol. 25 (2007) 1165–1170.
[37] J. Lu, Z. Li, J.I. Zink, F. Tamanoi, In vivo tumor suppression efficacy of mesoporous silica nanoparticles-based drug-delivery system: enhanced efficacy by folate modification, Nanomedicine 8 (2012) 212–220.
[38] R.D. Vinluan III, J. Zheng, Serum protein adsorption and excretion pathways of metal nanoparticles, Nanomedicine (Lond) 10 (2015) 2781–2794.
[39] Z. Chen, H. Chen, H. Meng, G. Xing, X. Gao, B. Sun, X. Shi, H. Yuan, C. Zhang, R. Liu, F. Zhao, Y. Zhao, X. Fang, Bio-distribution and metabolic paths of silica coated CdSeS quantum dots, Toxicol. Appl. Pharmacol. 1 (2008) 364–371.
[40] C. Valentina, C. Argyo, T. Bein, Impact of different PEGylation patterns on the long-term bio-stability of colloidal mesoporous silica nanoparticles, J. Mater. Chem. 20 (2010) 8693–8699.
[41] N. Feiner-Gracia, M. Beck, S. Pujals, S. Tosi, T. Mandal, C. Buske, M. Lindén, L. Albertazzi, Super-resolution microscopy unveils dynamic heterogeneities in nanoparticle protein corona. Small 13 (2017). https://doi.org/10.1002/smll.201701631.
[42] R. Michel, S. Pasche, M. Textor, D.G. Castner, The influence of PEG architecture on protein adsorption and conformation, Langmuir 21 (2005) 12327–12332.
[43] J. Croissant, X. Cattoën, M.W.C. Man, A. Gallud, L. Raehm, P. Trens, M. Maynadier, J.-O. Durand, Biodegradable ethylene-bis(propyl)disulfide-based periodic mesoporous organosilica nanorods and nanospheres for efficient in-vitro drug delivery, Adv. Mater. 26 (2014) 6174–6180.
[44] L. Maggini, I. Cabrera, A. Ruiz-Carretero, E.A. Prasetyanto, E. Robinet, L. De Cola, Breakable mesoporous silica nanoparticles for targeted drug delivery, Nanoscale 8 (2016) 7240–7247.
[45] X. Du, F. Kleitz, X. Li, H. Huang, X. Zhang, S.-Z. Qiao, Disulfide-bridged organosilica frameworks: designed, synthesis, redox-triggered biodegradation, and nanobiomedical applications, Adv. Funct. Mater. 115 (2018) 17325–17360.

CHAPTER EIGHT

Biodegradable Silica-Based Nanoparticles: Dissolution Kinetics and Selective Bond Cleavage

Jonas G. Croissant[*,†,1], **C. Jeffrey Brinker**[*,†]
[*]Chemical and Biological Engineering, University of New Mexico, Albuquerque, NM, United States
[†]Center for Micro-Engineered Materials, Advanced Materials Laboratory, University of New Mexico, Albuquerque, NM, United States
[1]Corresponding author: e-mail address: jonascroissant@unm.edu

Contents

Abstract

Silica-based nanomaterials are extensively used in industrial applications and academic biomedical research, thus properly assessing their toxicity and biodegradability is essential for their safe and effective formulation and use. Unfortunately, there is often a lot of confusion in the literature with respect to the toxicity and biodegradability of silica since various studies have yielded contradictory results. In this contribution, we first endeavor

The Enzymes, Volume 43
ISSN 1874-6047
https://doi.org/10.1016/bs.enz.2018.07.008

to underscore that the simplistic model of silica should be discarded in favor of a more realistic model recognizing that all silicas are not created equal and should thus be considered in the plural as silicas and silica hybrids, which indeed hold various biocompatibility and biodegradability profiles. We then demonstrated that all silicas are—as displayed in Nature—degradable in water by dissolution, as governed by the laws of kinetics. Lastly, we explore the vast potential of tuning the degradability of silica by materials design using various silica hybrids for redox-, pH-, enzymatic-, and biochelation-mediated lysis mechanisms.

1. INTRODUCTION: FROM SILICA TO SILICAS

1.1 From Silica to Silicas

It has been well said that all silicas are not created equal [1] and that we should no longer refer to silica but silicas [2]. The merit of these assertions is to clarify the fact that over-simplifying the properties of silica [3] is hindering the dissemination of knowledge in the literature. Typically, some published articles using silica particles asserted that silica is toxic, yet most of these studies failed to provide sufficient synthetic and characterization information to identify what kind of silica particles were examined. For instance, the size, aggregation state, charge, effectiveness of the surfactant removal (when applicable), surface functions, and radical-producing functions of the particles are all too often not characterized. Similarly, various publications dogmatically claimed that silica is not degradable, which we extensively refuted in a recent review article [4]. What these claims have in common is an over-simplification in regard to the nature of the silica structure and its properties, which in turn does not serve well the scientific and social communities (Fig. 1).

1.2 Biodegradability in Context

The terms degradability, biodegradability, and dissolution will be used interchangeably for the sake of simplicity, although the International Union of Pure and Applied Chemistry (IUPAC) recommended the use of the biodegradation terminology solely for a "degradation caused by enzymatic process resulting from the action of cells" [4,5].

Silica (SiO_2) is unstable in water and dissolves into silicic acid molecules ($Si(OH)_4$, pK_a 9.6) [4]. $Si(OH)_4$ is soluble in aqueous media and is the dominant species at low concentration ($<2 \times 10^{-3}$ M) [6], has an excellent

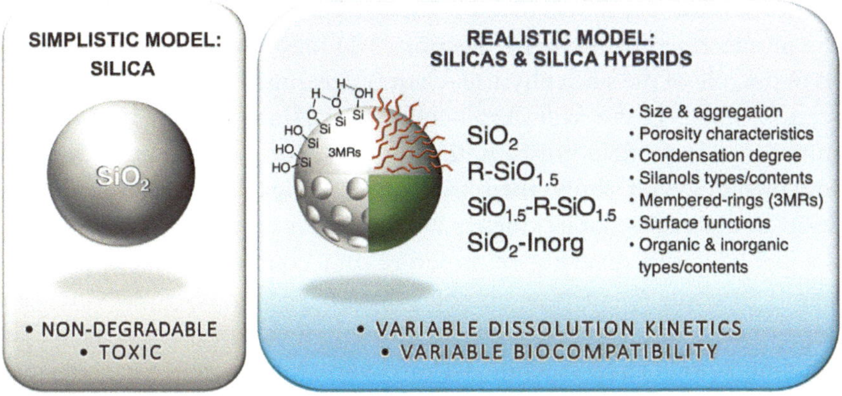

Fig. 1 Revisiting the current silica model with a more realistic model involving silicas and silica hybrids to account for the variability of the biodegradability rate and biocompatibility of silica particles as well as design biodegradable and biocompatible nanocarriers.

bioavailability while providing health benefits [7], and is excreted through the urine in both mice and humans [7]. The biocompatibility of silica and its degradation byproducts accounts for its generally recognized as safe (or GRAS) classification by the US Food and Drug Administration (FDA) for over 50 years [4]. As such, silica and silica particles are routinely used as additives in the food, pharmaceutical, and consumer industries.

The degradability of silica nanoparticles should be considered in the context of their applications, the regulatory requirements associated with a given application, as well as customer demands. According to the above factors, degradability will either be optional or indispensable (e.g., consumer product vs pharmaceuticals), beneficial or detrimental (e.g., eco-friendly products vs non-durable products). For instance, using degradable silica nanoparticle-based aerogel coatings could be preferred for transparent short-term surface protections, while non-degradable analogs would be essential for long-term superhydrophobic coatings [8–11] designed to repel water droplets under various weather conditions. On the other hand, the FDA requires the complete clearance of all injected contrast agents in a relatively short period of time [4], which implies that all pharmaceutical nanoformulations [12–19] must be degradable in the human body in a reasonable time-frame which will depend on their application.

In this chapter, we will first see that silicas and silica hybrids have unique distinguishing features which should be considered as a foundation for discussing their biodegradability (Sections 2.1–2.3). The dissolution kinetics of silicas per se will then be discussed in details, beginning with startling perspectives form Nature (Sections 3.1) followed by a systematic analysis of the role of the silica physical–chemical features on the dissolution rates (Section 3.2) [7]. We will chronicle silica parameters including (but not limited to) the particle porosity, the condensation degree, and the surface functionalities, and show their tremendous impact on the silica degradation under controllable kinetics ranging from days to months. We will also highlight the diversity of structure and content of silicas and silica hybrids (Section 4), such as silsesquioxane-doped or silsesquioxane-based nanoparticles, and how it widens the possible mechanisms of degradability including bioreduction (Section 4.1), pH variation (Section 4.2), enzymatic cleavage (Section 4.3), or biochelation (Section 4.4).

2. DISTINGUISHING FEATURES OF SILICAS

In this section, distinguishing physical–chemical features of silicas and silica hybrids are briefly overviewed (Fig. 2) to account for the variability of the degradation kinetics and mechanisms of the materials (see detailed discussion in Section 3).

2.1 Silica Nanoparticles

Silicon dioxide nanohydrates ($SiO_2 \cdot xH_2O$), better known as silica nanoparticles, are characterized by an inorganic network of tetrahedral SiO_4 units bonded via siloxane linkages (≡Si—O—Si≡) and bearing surface silanol groups (≡Si—OH) [20,21]. Silicas obtained at high temperatures such as fumed silicas have siloxane 3-members rings (3MRs [22]) which cause significant toxicity [1]. All silicas have various populations of several types of silanols groups: isolated silanols (≡Si—OH), geminal silanols (≡Si$(OH)_2$), and vicinal or H-bonded silanols (≡Si(OH)—O—(OH)Si═). If the silanols content of silica is very low, as a result of calcination for instance which can cause extensive silanol dehydration and siloxane formation (≡Si—OH + ≡Si—OH → ≡Si—O—Si≡ + H_2O), then the water dispersibility of silica will be very limited and its dissolution kinetics low. Silanols are deprotonated at pH 7 and form silanolates (or siloxide, ≡Si—O^-) due to the silica isoelectric point of ca. 2–3. All silanols, however, are not equal in regard to acidity since the pK_as of isolated, geminal, and H-bonded silanols are estimated to be 10.3, 2.9, and 7, respectively [23,24]. While the average density of silanols in amorphous silica is typically ~4.8 OH/nm^2, their density in various silicas differs

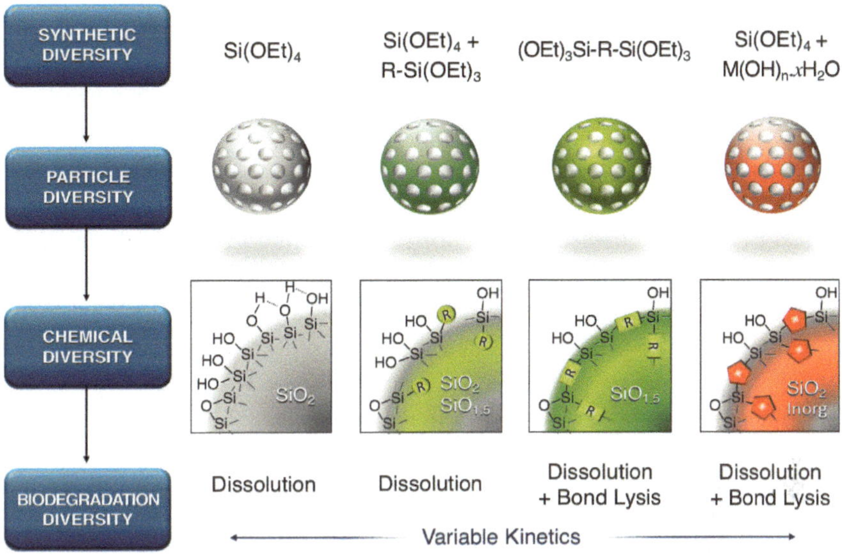

Fig. 2 Variability of the biodegradation kinetics according to the diversity of synthesis, particle features (nature, porosity, morphology, size), and the chemical structure and surface of silicas and silica hybrids (SiO_2, $SiO_{1.5}$, $SiO_2/SiO_{1.5}$, SiO_2/Inorg). R indicates organic groups; red pentagons indicate inorganic groups or nanodomains. From left to right, the particles represented are mesoporous silica, mesoporous organosilica, mesoporous silsesquioxane, and mesoporous silica-metal oxide nanoparticles.

significantly as follows: precipitated silica (~8.0 OH/nm^2) ≫ colloidal silica (~5.0 OH/nm^2) > microporous silica (4.6–4.9 OH/nm^2) > mesoporous silica (4.0–4.6 OH/nm^2) > silica gels (4.3–6.7 OH/nm^2) [25] ≫ commercial fumed silica (~2.0 OH/nm^2).

Silica particles also have varying levels of porosity. Stöber solid silica particles are microporous surface areas from 50 to 500 $m^2\ g^{-1}$[26–29], while the most common family of nanosilicas is mesoporous silica nanoparticles [7,20,30,31] or MSNs, which typically display surface areas from 600 to 1200 $m^2\ g^{-1}$[32–44]. Especially suited for therapeutic transportation [34,36,45–50], separation [37,51,52], and catalysis applications [53–60]. Unsurprisingly, the varying surface areas and surface properties of bare and surface-modified silica particles dramatically affect their dispersibility, biological interactions [7,61,62], and biodegradability rates [4] (see Section 3). Hence, even within the unity of the bare silica particle family, there is a great degree of structural and compositional diversity which translates into variable biological interaction and outcome [7].

2.2 Silsesquioxane Nanoparticles

Silsesquioxanes ($O_{1.5}Si\text{-}R$, with R organic groups) and organically bridged silsesquioxanes ($O_{1.5}Si\text{–}R\text{–}SiO_{1.5}$) [63] are essentially fully hybridized silica nanoparticles with exceptionally high organic content from 20 to 80 wt% [63–69]. The whole framework of silsesquioxane nanoparticles is constituted of organic groups either pending [63,70] from silicon atoms or bridging [68,71] silicon atoms. To date, organic pending groups include alkane, alkene, alkane-amine and large alkoxysilylated dyes [63], and organic bridging groups range from methylene ($\equiv Si—CH_2—Si\equiv$) [72], ethylene ($\equiv Si—CH_2—CH_2—Si\equiv$) [73], ethynelene ($\equiv Si—CH=CH—Si\equiv$) [74–77], phenylene ($\equiv Si—C_6H_4\text{-}Si\equiv$) [78–83], and biphenylene ($\equiv Si—C_6H_4—C_6H_4—Si\equiv$) [84–86] to more complex and larger conjugated groups such as tris(propyl)methylammonium [87], bipyridine [88–91], tetra(tetrazol-alkyl)porphyrins [69,73,92,93], tetra(tetrazol-alkyl) diphenylenebutadiaene [64,74], tetra(tetrazol-alkyl)phthalocyanine [94], to name a few. We refer the reader to several review articles for a more comprehensive treatment of silsesquioxanes nanomaterials [7,63,68,70,71, 95]. In terms of textural properties, designing mesoporous silsesquioxanes or periodic mesoporous organosilica nanoparticles (without silica source) has proved to be feasible with small-sized organic bridges (typically up to phenylene-bridged) but quite challenging with larger moieties (with the exception of a recent work [69]).

The full hybridization of silica with organic groups has many physical–chemical consequences including: (i) lower silanols contents (which for instance lowers the hemolytic activity of the particles [68,96]), (ii) a higher isoelectric point depending on the type of organic groups (ethylene-based PMO particles become positively charged at pH 5 [32,92,97]) which can be applied for autonomous pH-triggered drug delivery [32,78], (iii) higher thermal and mechanical stabilities [63,98,99], (iv) tunable hydrophobicity and hydrophilicity leading to exceptional cargo loading [73,78,98] and chromatographic capabilities [51,95,100,101], (v) electron-conductivity with conductive bridging groups [102], (vi) light-harvesting [103] via for instance biphenyl structural groups and coumarin guests [104], (vii) non-linear absorption with bridging groups stacking in favorable patterns [69], and (viii) stimuli-responsive biodegradability via the cleavage of specific bridging groups (Section 4). This non-exhaustive list is virtually unlimited since the properties of silsesquioxanes are dependent on the innumerable properties, contents, and arrangements of the organic groups.

2.3 Silica Hybrid Nanoparticles

Silica nanohybrids can also be prepared using silica sources with organic and/or inorganic dopants [4,69,105,106]. The typical strategy involves the preparation of organically doped silica particles by either condensing silica in the presence of organic doping agents or co-condensing silica with organosilane precursors. In the former case, the organic moieties are non-covalently trapped within the silica matrix, whereas in the latter case organic groups will be covalently linked to the framework via one ($O_{1.5}$Si-R) or more ($[O_{1.5}Si]_n$-R, $n \geq 1$) linkages. All doping strategies will impact the dissolution kinetics of the particles via increasing or lowering the probability of the cleavage of the silica network (vide infra). Alternatively, using alkali or metal complexes in conjunction with the silica source has been applied to dope silica with calcium, iron, zinc, and aluminum oxides. The homogeneity and type of the doping phase are key elements of such nanohybrids. The broad category of silica hybrid particles also includes silica particles engineered via surface modifications which totally redistribute the cards in terms of particle dispersibility, biological interactions, and dissolution kinetics. To summarize the distinguishing structural features of silicas, it should be kept in mind that the physical–chemical properties of silica dramatically vary with the synthesis method, post-modification, and overall design of the particles which in turn govern the kinetics of the silica dissolution.

3. BIODEGRADABLE SILICA: A QUESTION OF KINETICS

3.1 A Perspective From Nature

Silica is constantly forming and dissolving in Nature while silicon-containing molecules and materials are being transported by the elements. The overall world silica cycle is mainly correlated to weathering and biogenic uptake kinetics [107]. Considering the silicon cycle between the oceans and the shores of our planet, many transformational pathways for the transport of silicon involve the dissolution of silicas and silicates as well as the condensation of silicic acid molecules to form various types of silica-rich algae exoskeletons, marine glasses, clays, and minerals (Fig. 3). The biodegradability of silica is thus a constant reality in Nature and is it ruled by the laws of kinetics.

The terrestrial biosphere continuously feeds the oceans with silicon species. It is well known that the earth's crust consists in an estimated

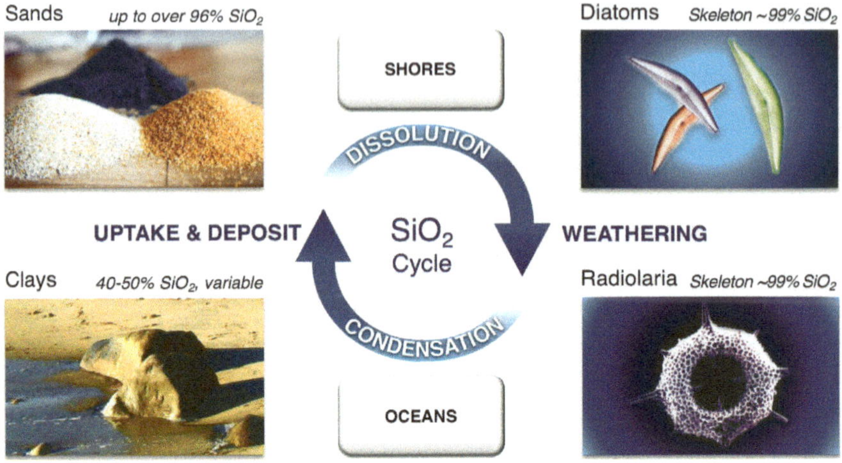

Fig. 3 Illustration of the natural silica cycle involving the dissolution and condensation of silica from the shores to the oceans. The formation of diatom and radiolarium skeletons takes place due to the presence of silicic acid molecules in the oceans. Silica skeletons eventually deposit on the ocean's floor (and are then dissolved) or on the seashores. Constant weathering processes continually feed the oceans with oligomeric silicas eroded for instance from sands and clays.

49% of silicate minerals, and as result, the weathering of the shores around the globe enriches the oceans in biogenic silica (9.4 Tmol per year, 6.2 Tmol originates from rivers) [108]. Notable sources of dissolved silica species include aluminosilicate clays (variable SiO_2 content, e.g., 40–50%), felsic rocks (45–69% SiO_2), and opals (>80–90% SiO_2) to name a few [109,110]. The thermodynamic rules of dissolution of silica are literally engraved in stones, even Martian stones as suggested by the discovery of high-purity opaline silica on Mars (up to 91% SiO_2) [111]. In addition to silica-rich clays and rocks on the earth, the great variety of sands (up to 96% of SiO_2), which is the principal silicon source for the electronic and glass industry, also contributes to biogenic silica in the oceans. Silicon-containing molecular species also come from human wastes and byproducts, including silicon-rich fertilizers (beneficial for plant fitness [112]), which are biodegraded and flow partly into the rivers and then end into the oceans.

The oceans, seas, and rivers hence contain various siliceous species [113,114]. The silica concentration in natural waters ranges from 1 to $30\,mL\,L^{-1}$ on average, while its aqueous solubility limit approximates $120\,mg\,L^{-1}$ at 25°C [115]. The main characters in the silicon cycle of the oceans are marine radiolaria and diatoms which are silica-precipitating microalgae. Such species interestingly account for well over half of the organic carbon accumulation in ocean sediments, and as such, significantly impact the global carbon cycle [108]. As noted above, the continental erosion provides the ocean with bioavailable silicic acid which is then utilized by radiolaria and diatoms to form their magnificent porous silica exoskeletons (Fig. 3, right panel). The weathering processes have in fact been correlated to the increased population of marine diatoms over time [108]. Radiolarium and diatom skeletons eventually deposit onto the ocean's floor and are dissolved [116]; a mechanism which can notably be accelerated by marine bacterial assemblages [117]. In fact, bacteria, fungi, worms as well as cells and tissues, all catalyze the dissolution of both amorphous and crystalline silicas [118]. Diatoms alone precipitate approximately 240 Tmol of biogenic silica every year [113], which is amazingly contrasted with only 9.4 Tmol of biogenic silica eroded from the terrestrial biosphere each year. Even though it is likely that the extent of the erosion of silica is still underestimated, it is obvious that silica-rich marine species dominate the silica cycle in our era. In regard to the dissolution kinetics of biogenic silica, over half of diatomaceous silica produced on the surface of the ocean's water (the "euphotic zone" where light penetrates and photosynthesis occurs) redissolves within 100 m of the surface which contrasts with order-of-magnitude lower dissolution rates in deep-sea sediments [114]. This can be explained by the lower temperatures in the deeper oceans as well as the different compositions (Al/Si ratio) and structures (condensation degrees and surface areas) of the deposited biosiliceous oozes of the Southern Ocean [114]. Diatomaceous earth silica is thus an astonishing example of the amazing travel of silicon atoms on the earth from the condensation of silicic acid in the oceans to the sedimentation of silica-rich diatoms in rivers, lakes, and oceans, and then the collection of diatomaceous earth by the hundreds of tons worldwide every year for applications as explosives (Alfred Nobel stabilized nitroglycerin absorbed in diatomite and called it dynamite), adsorbents (high porosity), pest controls (insecticide property), and agricultural grain storage (anticaking agent property).

Silicas thus travels from shore to shore and is found in the oceans in the form of biogenic silica and dissolved silica. Similarly to biogenic and lithogenic silicas, the dissolution of silica nanoparticles applied for biomedical application occurs with variable kinetics, as detailed in the next section.

3.2 Tuning the Dissolution Kinetics

The dissolution of silica involves three chemical steps: (1) *hydration*: water molecules adsorb onto the silica surface ($\equiv$Si—O—Si$(OH)_3$ + H_2O) and hydrogen-bond with silanols [119], (2) *hydrolysis*: the nucleophilic attack of hydroxyl groups on silicon atoms ($\equiv$Si—O—Si$(OH)_3$ + $[HO^-;H^+]$ $\rightarrow$ $\equiv$Si—O—$Si^-(OH)_4^{\neq}$ + H^+), and (3) *ion exchange*: release of silicic acid molecules in the presence of water molecules ($\equiv$Si—O—$Si^-(OH)_4^{\neq}$ +H^+ + $[HO^-;H^+]$ $\rightarrow$ $\equiv$Si—OH + $Si(OH)_4$ + H_2O). Notably, using aqueous media are critical to degrade silicas, and for instance, the dissolution of silica particles can be virtually inhibited in absolute ethanol [30]. Silica-based nanoformulations may be stably stored in absolute ethanol for a couple of years, as far as silica is concerned, but in water media, the particles can degrade as fast as 1 day.

Tuning the aqueous dissolution kinetics of silicas can be achieved by first engineering the physical–chemical properties of the materials [4,120] (Fig. 4A) including the following: (i) the surface area, (ii) the condensation degree, (iii) the type and content of silanol groups, (iv) the presence of surface capping agents (R-$SiO_{1.5}$), (v) the aggregation state, and (vi) the size and morphology of the particles. The dissolution kinetics of silica in water secondly depends the media parameters [4,118,121,122]: (i) the pH, (ii) the temperature, (iii) the particle concentration, (iv) the protein content, and (v) the type of system (i.e., static/continuous flow; in vitro/in vivo). Selected parameters will now be discussed in greater details; for an extensive treatment on the subject, we refer the readers to our recent comprehensive review [4].

First, the higher the silica surface area is, the higher the dissolution kinetics. The surface plays a key role in the dissolution kinetics since it is correlated with the accessibility of silicon centers where the dissolution process is initiated. In addition, higher surface areas accelerate the dissolution kinetics due to the increased diffusion of silicic acid byproducts (the degradation is correlated with the solubility limit of silicic acid in a given nanoenvironment). Another factor interplaying with the higher surface areas is the associated lower condensation degrees (vide infra). As a result, when comparing solid silica spheres and mesoporous nanoparticles, Kuroda et al. observed than non-mesoporous particles were fully degraded after about 12 days while mesoporous particles did so in only 6 days [96]. Multiples studies have reported the aqueous biodegradation of MSNs with high surface areas in 7 days or less [123–126]. Aggregating 20 and 30 nm-wide MSNs also

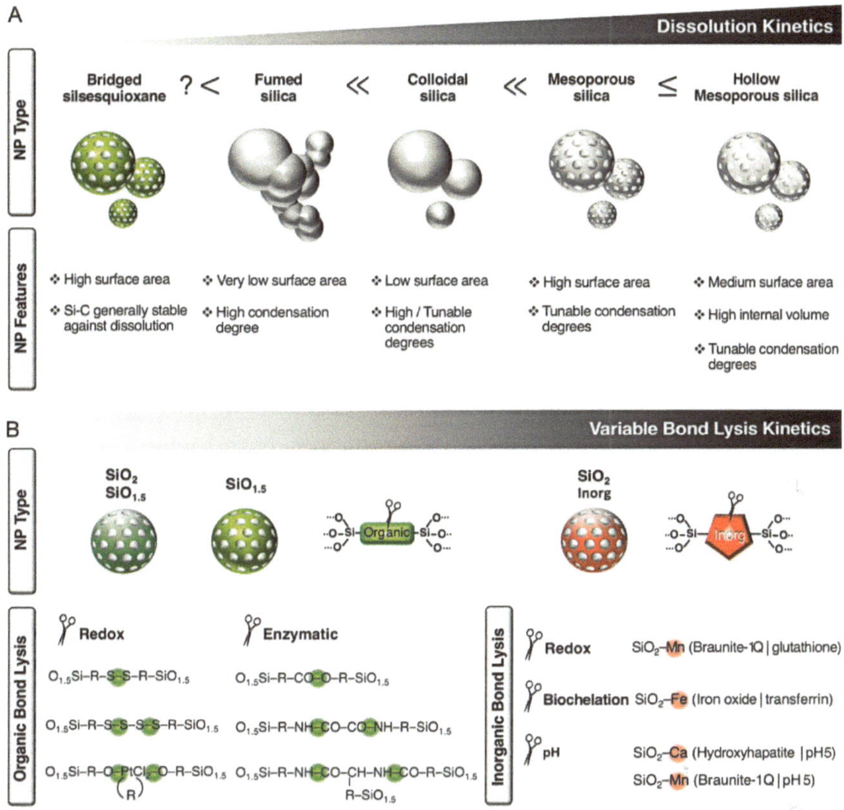

Fig. 4 Representation of the silica dissolution rate (A) and bond lysis kinetics (B) variations through nanomaterials engineering of silicas. Silica networks are unstable in water and thus dissolve with variable kinetics. Organic or inorganic incorporation with siliceous networks enable additional degradation mechanisms in addition to dissolution.

changed their dissolution curves from exponential to linear rates. Based on these results, the dissolution kinetics of fumed silica should be lower than colloidal silica analogs (Fig. 4A). Increasing the size of the solid silica particles also decreases their dissolution kinetics (all other parameters being equal) [96] since it implies lower external surface areas and longer diffusion rates of dissolved species. Shi and co-workers confirmed the major role of the porosity on the degradation of particles possessing surface areas of 958, 829, and 282 $m^2\ g^{-1}$ and displayed burst dissolutions in the first 2–4 h with 30%, 70%, and 90% of degradation, respectively, and total degradations after 2 weeks [127].

Secondly, the lower the silica condensation degree is, the higher the dissolution rate. The condensation degree of silica is the Q^3/Q^4 ratio where Q^n stands for silicon atoms bearing four oxygen atoms and 4-*n* OH groups. This population ratio in silicas materials can be determined by quantitative solid-state ^{29}Si nuclear magnetic resonance (NMR) spectroscopic methods such as proton polarization decoupling (HPDEC) sequences. The condensation degree is indeed a key parameter regulating the biodegradability of silica particles as illustrated by the dramatic difference of degradation kinetics ranging from 1 day to several weeks or months [4]. Specifically, Shen and coworker showed that low condensation degrees ($Q^3/Q^4 \sim 0.4$–0.5) lead to extremely fast degradation of 100 nm-wide MSNs in 1 or 2 days, whereas higher degrees ($Q^3/Q^4 \sim 0.1$–0.2) significantly lowered the dissolution to 4 weeks [126]. This can be explained by the impact of higher the condensation degrees, which require more hydrolysis reactions to release $Si(OH)_4$ moieties.

In addition, in the case of silicas which synthetic history involves for instance high temperature and pH treatments, the content of siloxanes can be so high as to strongly limit the dispersibility of the particles in water, hence further decreasing the silica dissolution rate. Shi et al. showed that only 30% of calcined silicas dissolved in 2 weeks in SBF medium whereas partially condensed silicas had fully degraded [127]. Finnie et al. observed that nonporous silica nanoparticles degrade much slower than mesoporous analogs, as explained by their higher condensation degrees ($Q^3/Q^4 \sim 0.2$) [128]. Lastly, gradient or different in condensation degree in one particle has been shown to produce intermediate hollow mesoporous particles from MSNs having less condensed cores [129,130]. In fact, the same principle is often used to produce well-defined hollow MSNs with the selective etching of the hard silica nanoparticle template [131].

Thirdly, it was established that the dissolution rate (R_{diss}, see equation hereafter) of silica materials depends on the type and contents of silanol groups [118,132]: (1) protonated ($\equiv SiOH_2^+$), (2) neutral ($\equiv SiOH$), and (3) deprotonated silanols ($\equiv SiO^-$):

$$R_{diss} = k(H^+)[SiOH_2^+]^m + k(H_2O)[SiOH]^m + k(HO^-)[SiO^-]^p$$

where *m* and *p* are reaction orders; $k(H^+)$, $k(H_2O)$, and $k(HO^-)$ are rate constants.

As highlighted in Section 2.1, silicas obtained under different synthetic processes display variable surface silanol densities ranging from 2 to

8 OH/nm^2 and silanol of various kinds have a pK_a ranging from 2.9 to 10.3. As a result, the probability of the hydrolysis of silica into silicic acid can vary significantly according to the type and content of silanol groups as well as the pH. To the best of our knowledge, systematic degradation studies using silica nanoparticles with quantified silanol populations have not yet been reported, even though the role of silanol types and 3MRs have been linked to silica toxicity [1,2,133]. This is thus an area of research requiring systematic investigations.

Silanol populations are pH dependent and so will be the silica dissolution. Analyzing 200 nm-wide MSNs, Chen et al. observed a minimal degradation at pH 4.7 (acetate buffer) after 2 days, as opposed to the extensive degradation obtained at pH 7.4 (PBS buffer) [130]. The group of Lindén described that both MSNs and nonporous silica particles dissolved in the following order: simulated lung fluid (pH 7.4) > simulated body fluid (pH 7.25) ≈ PBS (pH 7.4) ≫ simulated gastric juice (pH 1.6) [122]. The effect of the pH was clear, along with that of the presence of organic acid salts in the buffers, such as di- and tri-carboxylates which prevent silica precipitation by forming silicates complexes in water and act as nucleophiles on positively charged silicon centers [134,135]. The authors also confirmed the role of the sodium ions in enhancing the deprotonation of the surface of silica particles [122]. The above-mentioned trends thus highlight that the pH, as other parameters, cannot be taken as a stand-alone parameter to predict the degradability rate of silica. The whole degradation medium should be considered including the parameters such as the ionic strength, alkali ions, proteins and organic acid salts, temperature, as well as the particle concentration and the type of system (continuous flow vs static). Importantly, the classical model of the synthesis and stability of silica sols [136,137] over various pHs (which states that particles only dissolve above pH 10) should hence not be uncritically considered for the post-synthesis stability analysis of the sols since this model typically involves concentrated solid silica (micro) particles in pure water and static conditions.

Fourthly, the covalent surface functionalization of silica particles can act in various capacities as a passivation layer against a rapid degradation in water [138–140]. At least 15 days of stability against dissolution was reported by the group of Haynes via functionalizing polyethylene glycol (PEG) polymers onto MSNs [140]. Investigating MSNs pegylated in the three different ways (1: Mw = 550 Da; 2: Mw = 5000 Da; 3: Mw = 550 + 5000 Da) and then soaked the particles in SBF media for a month [141], Cauda et al. reported that longer and denser PEG grafting lowered the dissolution rates.

The dissolution kinetics of MSNs in FBS at 37°C was effectively decreased without changing the final degradation percentage through pegylation [140], while grafting polyethyleneimine (PEI) polymers both decreased the kinetics and the extent of degradation by 25% in PBS at 37°C [139]. The passivation of the silica surface against dissolution via co-grafting both PEG and PEI lead to the dissolution of MSNs from the inside out [4]. Finally, the covalent surface engineering of various silicas with silsesquioxanes bearing aliphatic carbon chains, amine, carboxylate, or phenyl groups generally lead to lower dissolution kinetics and/or efficacies in various studies [4]. Note that the non-covalent doping of silicas, such as loading cargo molecules, was also reported to modify the biodegradation kinetics of silica nanocarriers, either increasing its extent [142], and accelerating [4] or retarding its kinetics [143]. The counterintuitive increased kinetics and/or extent of the dissolution processed can be explained in part by lower condensation degree is organically doped silicas [4].

The higher aqueous stability of Si—C bonds when compared to Si—O bonds accounts for the absence of the dissolution of bridged silsesquioxane and PMO nanomaterials in current short-term (few days to 1 month) degradation studies (Fig. 4A) [4]. Particles composed of ethylene (≡Si—CH_2—CH_2—Si≡) [97], ethynelene (≡Si—CH=CH—Si≡) [32,96], phenylene [144], and curcumin [97] bridges were all showed to resist dissolution. Studies remain to be performed to ascertain the maximum stability of these structures in aqueous media. The degradation of MSNs, BS, and PMO NPs can, however, be achieved by other mechanisms by doping the particles with organic or inorganic removable components as discussed in the next section.

4. BIODEGRADABLE SILICA HYBRIDS: USING SELECTIVE BOND CLEAVAGE

Silicas and silica hybrids such as silsesquioxanes and silica-mixed oxides can be degraded by other mechanisms in addition to dissolution in water via organic or inorganic doping (Fig. 4B). Doping the particle framework with these approaches enables the controllable particle biodegradation using redox-, enzymatic-, pH-, and biochelation-mediated bond cleavages.

4.1 Redox-Mediated Lysis

Several redox-sensitive bridging groups were incorporated into silicas and silica hybrids including disulfides [32,47,66,145–153], tetrasulfides

[154–160], and cis-platin bridges [161,162]. The former two functional groups are the most widely used for their capability to generate an autonomous remote degradation of the nanoparticles capitalizing on the presence of bioreducers inside cells. Typically, glutathione reducers are present in much higher concentrations ($2–10 \times 10^{-3}$ M) in cancer cells than in healthy cells ($2–20 \times 10^{-6}$ M) and predispose nanoformulations to degrade preferentially inside tumors. Silica nanoparticles were synthetized by co-condensing tetraethoxysilane with bis(triethoxysilylpropyl)disulfide by Quignard and co-workers (up to 40 wt% in disulfide) [153]. They described decreased dissolution kinetics with higher disulfide contents which were attributed to an increased framework hydrophobicity. Redox-mediated biodegradation experiments were then successfully performed at 37°C with dithiothreitol model reducers (1.3×10^{-3} M), leading to extensive particle fragmentation and dissolution after 7 days. Similarly, Zhang et al. synthesized and degraded in 5 days 200 nm-wide hollow nonporous disulfide-bridged (11 wt%) organosilica particles by immersion in PBS with glutathione (8×10^{-3} M) [146]. Croissant et al. prepared solid disulfide-photosensitizer-bridged silsesquioxane nanospheres (70–90 wt% in disulfide) from 200 to 30 nm in diameter and demonstrated their biodegradation in only 48 h in PBS containing mercaptoethanol (2 mM) at 37°C. The biodegradable nanovectors were also applied for effective two-photo-triggered photodynamic therapy and bioimaging [66]. The same group also developed biodegradable ethylene-bis(propyl)disulfide-based PMO nanospheres and nanorods (10–50 wt% in disulfide) for autonomous drug delivery in cancer cells [32]. The particles did not dissolve in the control experiments and fragmented by adding mercaptoethanol bioreducers both at 6 μM and 2 mM (Fig. 5). The unique built-in hydrophobicity/hydrophilicity balance of the mesopores provided a redox-mediated biodegradation, high water dispersibility and drug loading (25 wt%), virtually zero passive leakage, as well as a pH-triggered delivery by the particle charge reversal at pH 5 with ~90% cell killing at particle concentration of only 1 μg/mL [32]. Another recent publication systematically comparing MSNs with disulfide-doped MSNs showed enhanced DOX delivery and anti-tumor efficacy in cancer cells of the latter [163]. Importantly, the degradation of disulfide-doped hollow mesoporous silica nanocomposites effectively occurs within cells [4] after accumulation in the endolysosomal compartments [164]. Importantly, Tamanoi and co-workers demonstrated that DOX-loaded tetrasulfide- and disulfide-doped PMO nanospheres have higher accumulations and tumor growth inhibitions than mesoporous silica particles [165].

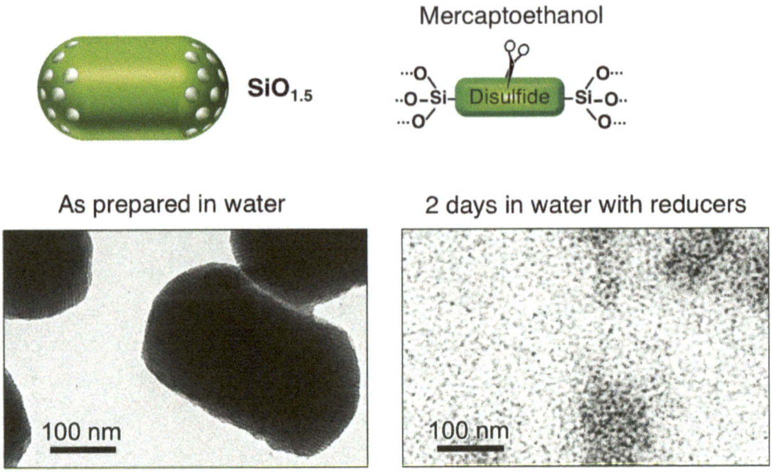

Fig. 5 Representation of the redox-mediated degradation of mesoporous ethylene-bis(propyl)disulfide-based silsesquioxane nanoparticles along with transmission electron microscopy images of the particles before and after addition of mercaptoethanol reducers (2 mM). The particles have a hexagonally ordered mesoporosity before degradation. *Unpublished data related to the published work of J. Croissant, X. Cattoën, M.W.C. Man, A. Gallud, L. Raehm, P. Trens, M. Maynadier, J.O. Durand, Biodegradable ethylene-bis (propyl) disulfide-based periodic mesoporous organosilica nanorods and nanospheres for efficient in-vitro drug delivery, Adv. Mater. 26 (2014) 6174–6180.*

cis-Platin and disulfide groups were also incorporated into bridged silsesquioxane nanomaterials to function as pro-drug nanotherapeutics. The group of Lin took advantage of the redox-cleavable property of cis-platin-bridges constituting BS nanotherapeutics to delivery cis-platin drugs [161,162]. Using cysteine reducing agents in aqueous solution (10×10^{-3} M), the pegylated nanocarriers were first shown to controllably release their content and were secondly applied for cis-platin delivery to bind DNA and inhibit tumor growth. Anisamide-functionalized analogs were also used in a pancreatic cancer xenograft mouse model to effectively inhibit tumor growth when compared to free oxaliplatin [162]. Vivero-Escoto and co-workers prepared biodegradable BS NPs bearing protoporphyrin IX and disulfide bridges for photodynamic therapy application via the pro-drug delivery of the photosensitizer [166]. Gadolinium-disulfide-bridged BS NPs were also effective magnetic resonance imaging (MRI) contrast agents using a similar approach [167].

In closing the section of redox-mediated degradation of mesoporous silica particles, it is noteworthy that copper [168], manganese [169], and calcium-doped [170] silica nanocomposites were also shown to be redox sensitive although their degradation mechanisms remain to be elucidated.

4.2 pH-Mediated Lysis

The pH-mediated bond lysis approach to degrade organosilica nanoparticles has been engineered by doping calcium and manganese oxides within the mesoporous silica nanostructure [4]. The leaching of calcium species from calcium-doped silica particles was observed to increase the hydrolytic degradability of the framework [170–177]. Dissolving calcium centers acted as hole generators within the silica framework which accelerated the nucleophilic attacks on silicon atoms and subsequent silica dissolution into silicic acid molecules. This approach was used for the sustained release of cargos such as ibuprofen. Mesoporous silica–hydroxyapatite nanocomposites were also prepared with a hexagonal $Ca_{10}(PO_4)_6(OH)_2$ (hydroxyapatite) component [178] with Si–O–Ca–O–Si linkages [170]. The enhanced anticancer drug delivery was monitored under acidic conditions and associated with the release of degraded calcium centers which was a 100-fold faster at 5.0 than at pH 7.4 (Fig. 6A). Electron microscopy images demonstrated the fragmentation of the particles at pH 5.0. Another study showed that the degradation rates and maximum percentages of SiO_2–Ca correlated the variation of the pH from 7.4 to 4.5, the latter pH leading to a complete degradation after only 24 h [164].

Similarly to calcium oxides, manganese oxides are biocompatible and part and parcel of the fabric of the human body, which allows the safe

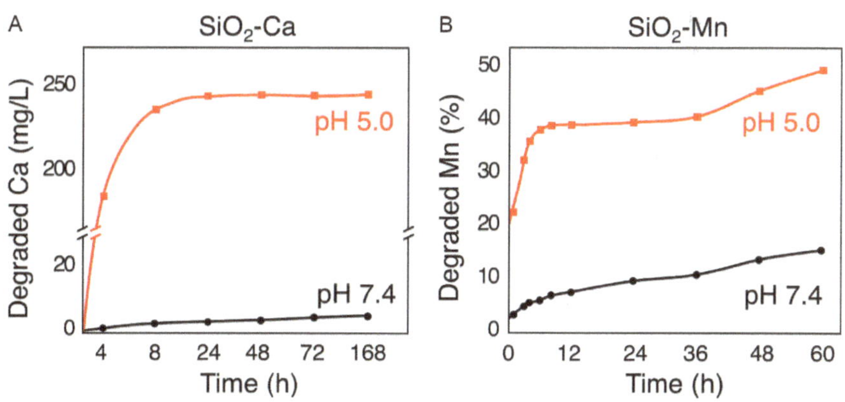

Fig. 6 Degradation kinetics of pH-sensitive Ca-doped (A) and Mn-doped MSN (B) at pH 5.0 and 7.4 demonstrating the impact of the acidity on the particle dissolutions. *Unpublished scheme based on the published works of X. Hao, X. Hu, C. Zhang, S. Chen, Z. Li, X. Yang, H. Liu, G. Jia, D. Liu, K. Ge, Hybrid mesoporous silica-based drug carrier nanostructures with improved degradability by hydroxyapatite, ACS Nano 9 (2015) 9614–9625; L. Yu, Y. Chen, M. Wu, X. Cai, H. Yao, L. Zhang, H. Chen, J. Shi, "Manganese extraction" strategy enables tumor-sensitive biodegradability and theranostics of nanoparticles, J. Am. Chem. Soc. 138 (2016) 9881–9894.*

use and excretion of SiO_2–Mn and SiO_2–Ca nanomaterials. The doping of manganese oxide into mesoporous silica particles was performed to trigger the biodegradation of the particles inside tumor microenvironments [169]. Nearly 10wt% of a braunite-1Q phase ($Mn^{2+}Mn^{3+}{}_6SiO_{12}$) was incorporated into sub-100nm hollow MSNs and afforded higher degradation kinetics of SiO_2–Mn at pH 5.0 than at physiological pH (Fig. 6B). Thanks to the magnetic property of the manganese doping, SiO_2–Mn nanocarriers were loaded with DOX drugs for in vivo cancer therapy and MRI tumor imaging. In addition to being pH sensitive, SiO_2–Mn and SiO_2–Ca are also redox sensitive, hence affording pH- and glutathione-actuated synergistic biodegradation mechanisms. The fragmentation of both particle types into small-sized particulates allowed their rapid excretion in the urine and feces of mice after effective therapeutic applications [4].

4.3 Enzymatic Mediated Lysis

Another promising approach to control the biodegradation rate of silica nanoparticles is to engineer them with built-in protein or enzyme-cleavable groups. A number of surface-engineered silica and organosilica particles have been designed for drug delivery in the presence of enzymes [179–185], but engineering the framework of silica nanohybrids is the approach which allows an enzyme-mediated lysis and the subsequent degradation of nanomaterials [186]. The first example to our knowledge of using enzymatically degradable groups into a nanoformulation used an organosilica-liposome nanohybrid designed to load and deliver DOX drugs. Liposomes were stabilized by a 20nm-thick organosilica layer consisting of an ester-silica structure, presumably enzymatically responsive to account for the autonomous intracellular delivery of the cargos [187]. This approach was pursued to take advantage of the liposome assets such as biocompatibility and high drug loading capacity while enhancing their stability and controlled delivery via an organosilica hybridization.

Organosilica robust nanocarriers were then designed to biodegrade in cells in the presence of enzymes to minimize the passive leakage of therapeutic cargos. A proof of concept of this functionality was reported by Fatieiev and co-workers using a tailor-made amino-acid-like bridge silsesquioxane precursor bearing an oxamide function [67], which was later incorporated into a ca. 100nm-wide mesoporous hybrid silsesquioxane nanoparticle for trypsin enzyme-mediated degradation in 3 days (Fig. 7) [78]. This work described the first inorganic nanoparticle capable of degradation through

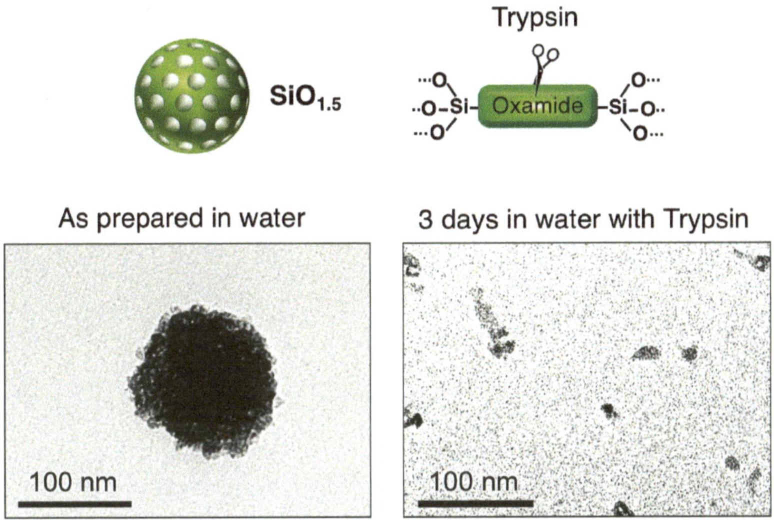

Fig. 7 Representation of the enzyme-mediated degradation of bis(alkyl)oxamide-bridged silsesquioxane nanoparticles along with transmission electron microscopy images of the particles before and after the addition of trypsin enzymes. These particles have a disordered worm-like mesoporosity. *Unpublished data related to the published work of J. Croissant, Y. Fatieiev, K. Julfakyan, J. Lu, A. Emwas, D. Anjum, H. Omar, F. Tamanoi, J. Zink, N. Khashab, Biodegradable oxamide-phenylene-based mesoporous organosilica nanoparticles with unprecedented drug payloads for delivery in cells, Chem. Eur. J. 22 (2016) 14806–14811.*

enzymatic cleavage and also displayed an unprecedented high loading of hydrophilic and hydrophilic drugs (up to 84 wt%) with a pH-triggered delivery (~pH 5.5) in cancer cells. These features arose from the remarkably high organic content of the particle (~50 wt%, including 24 wt% of cleavable groups) while maintaining a higher surface area of 850 $m^2 g^{-1}$. Unlike silica nanoparticles which require surface mechanization to prevent the leakage of cargo molecules [36,49,188–190], these biodegradable mesoporous silsesquioxane nanocarriers did not require this precaution and hosted important amounts of drugs without leakage which is cost-effective and removes the risks associated with the potential toxicity of degraded byproducts of the molecular mechanization.

Another original enzymatically degradable non-mesoporous donut-shaped nanoparticle was invented by De Cola et al. using a tetrasilylated lysine-bridged silsesquioxanes with as high as 70% of peptide content [191]. The biodegradation was dependent on the concentration of trypsin peptidase enzymes (as demonstrated by electron microscopy images) for the delivery of DOX drugs in vitro. The donut-shaped carrier was degraded in solution after 3 days, and the degradation was also visualized over time in HeLa Cells. Peptidase enzymatically degradable nanomaterials are the next generation of nanotherapeutics for the selective delivery of drugs into specific organs for disease-targeted treatments [78]. This represents a paradigm shift in the field of spatio-controlled nanomedicine since the performance, selectivity, and fate of the nanocarrier are then governed by materials design [7,78].

4.4 Biochelation-Mediated Lysis

In order to implement the biochelation-actuated biodegradation of silica-based particles, the doping of bio-responsive metal oxides was investigated, namely various polymorphs of iron oxide-silica nanocomposites (e.g., with FeOOH, Fe_3O_4, Fe_2O_3 phases) [4,192]. Similarly to the doping of calcium and manganese, higher dissolution rates arise in SiO_2–Fe as a result of two factors, namely lower condensation degrees of the silica at the heterogeneous oxide-silica interfaces, and the generation of porosity upon the iron oxide removal [193–196]. Hollow Fe-doped MSN (~4.1 M% of Fe) of 200 nm of diameters displayed high dissolution rates upon the addition of different chelating agents (e.g., ethylenediaminetetraacetic acid) by removing the nearly 40% of iron oxide nanophases after 24 h [193]. The nanoparticles rapidly dissolved in human fetal bovine sera in 20–25 days while hollow silica particle controls did not. Similar biodegradations in protein-containing media were observed for α-FeOOH–SiO_2 (~19 wt% of iron oxide hydroxide) [195] and Fe_2O_3–SiO_2 (~48 wt% of iron oxide) [197] and the key role of transferrin proteins in removing the iron centers, due to their high binding constant of about 10^{20}, was identified [197]. The latter Fe_2O_3–SiO_2 nanoparticles having 20–50 nm-wide pores, were fully degraded in 3 days when dispersed in transferrin-rich fetal bovine serum, as shown by electron microscopy analysis (Fig. 8) [197]. In addition to providing enhanced biodegradation in controlled environments, the doping of antiferromagnetic or superparamagnetic iron oxide afforded in vivo T_1 MRI [194], magnetically triggered release of ultra-large proteins [197], and biocatalysis [192], thus demonstrating that silica nanocomposites are not only advantageous for their enhanced and tunable biodegradability but also for their multifunction integration.

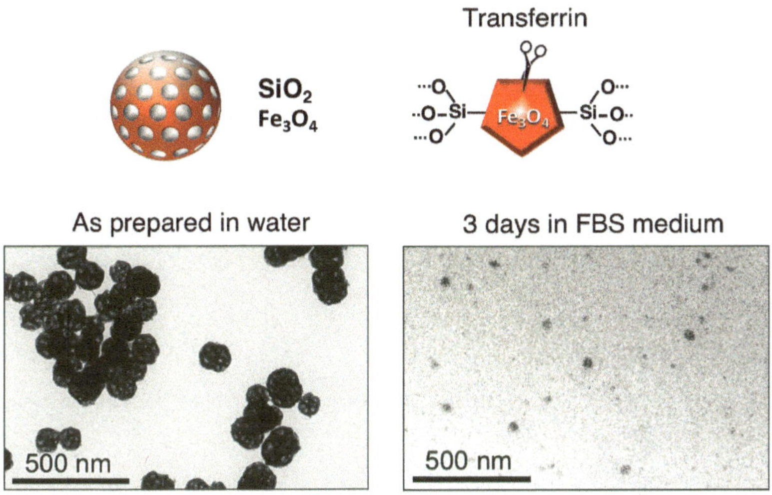

Fig. 8 Representation of the biochelation-mediated degradation of silica–iron oxide nanoparticles along with transmission electron microscopy images of the particles before and after the addition of transferrin chelating agents. These particles have a disordered cavity-like mesoporosity. *Unpublished data related to the published work of H. Omar, J.G. Croissant, K. Alamoudi, S. Alsaiari, I. Alradwan, M.A. Majrashi, D.H. Anjum, P. Martins, B. Moosa, A. Almalik, N.M. Khashab, Biodegradable magnetic silica@iron oxide nanovectors with ultra-large mesopores for high protein loading, magnetothermal release, and delivery, J. Control. Release 259 (2017) 187–194.*

5. OUTLOOK

This chapter first started with the assertion that all silica is not created equal, hence implying that it more scientifically precise to refer to silicas rather than silica since not all silicas are equality biocompatible and biodegradable. There is indeed a vast diversity of structure and properties within the unity of the SiO_2 formula. The structural diversity, including varied silanols and organic contents, condensation degrees, porosity, and surface chemistry, accounts for the dissolution rate properties varying from 1 day to several months. Such an observation is nothing new since natural silicas found everywhere on earth and in the oceans, such as in radiolarians and diatoms, has been forming and dissolving for eons. What is new, is not the tuning of the silica dissolution kinetics, though it has been harnessed for the design of silica nanotherapeutics better suited for an FDA approval, the novelty is the use of specific bond lysis within the silica framework for selective endogenous biodegradations. The state of the art research on the biodegradability of silica nanoparticles has thus focused on fabricating redox-, pH-, enzymatic-, and biochelation-bioresponsive nanoparticles. These features were implemented via the organic and inorganic doping of silica or silsesquioxane nanomaterials. Recent research has shown that such nanohybrids often display superior loading capacity, drug delivery efficacy, and biodegradability rates and selectivity than pure silica, which suggests that the search for new biodegradable silica hybrids is a rewarding research field to pursue.

ACKNOWLEDGMENTS

Jonas G. Croissant and C. Jeffrey Brinker acknowledge support by the University of New Mexico for this work.

REFERENCES

[1] H. Zhang, D.R. Dunphy, X. Jiang, H. Meng, B. Sun, D. Tarn, M. Xue, X. Wang, S. Lin, Z. Ji, R. Li, F.L. Garcia, J. Yang, M.L. Kirk, T. Xia, J.I. Zink, A. Nel, C.J. Brinker, Processing pathway dependence of amorphous silica nanoparticle toxicity: colloidal vs pyrolytic, J. Am. Chem. Soc. 134 (2012) 15790–15804.
[2] C. Pavan, B. Fubini, Unveiling the Variability of "quartz hazard" in light of recent toxicological findings, Chem. Res. Toxicol. 30 (2016) 469–485.
[3] C.J. Brinker, G.W. Scherer, Sol–Gel Science: The Physics and Chemistry of Sol–Gel Processing, Academic Press, 2013.
[4] J.G. Croissant, Y. Fatieiev, N.M. Khashab, Degradability and clearance of silicon, organosilica, silsesquioxane, silica mixed oxide, and mesoporous silica nanoparticles, Adv. Mater. 29 (2017) 1604634.
[5] M. Vert, K.-H. Hellwich, M. Hess, P. Hodge, P. Kubisa, M. Rinaudo, F. Schué, Terminology for biorelated polymers and applications (IUPAC recommendations 2012), Pure Appl. Chem. 84 (2012) 377–410.
[6] K.R. Martin, Silicon: the health benefits of a metalloid, in: Interrelations Between Essential Metal Ions and Human Diseases, Springer, 2013, pp. 451–473.
[7] J.G. Croissant, Y. Fatieiev, A. Almalik, N.M. Khashab, Mesoporous silica and organosilica nanoparticles: physical chemistry, biosafety, delivery strategies, and biomedical applications, Adv. Healthcare Mater. 7 (2018) 1700831.
[8] S. Singh, J. Houston, F. van Swol, C.J. Brinker, Superhydrophobicity: drying transition of confined water, Nature 442 (2006) 526.
[9] D.A. Doshi, P.B. Shah, S. Singh, E.D. Branson, A.P. Malanoski, E.B. Watkins, J. Majewski, F. van Swol, C.J. Brinker, Investigating the interface of superhydrophobic surfaces in contact with water, Langmuir 21 (2005) 7805–7811.
[10] E.D. Branson, P.B. Shah, S. Singh, C.J. Brinker, Preparation of hydrophobic coatings, Google Patents, 2009.
[11] H.R. Fraval, C.J. Brinker, Fluid light guide having hydrophobic aerogel cladding layer, Google Patents, 2006.
[12] G. Bozzuto, A. Molinari, Liposomes as nanomedical devices, Int. J. Nanomedicine 10 (2015) 975–999.
[13] R. van der Meel, T. Lammers, W.E. Hennink, Cancer Nanomedicines: Oversold or Underappreciated? Taylor & Francis, 2017.

[14] D. Bobo, K.J. Robinson, J. Islam, K.J. Thurecht, S.R. Corrie, Nanoparticle-based medicines: a review of FDA-approved materials and clinical trials to date, Pharm. Res. 33 (2016) 2373–2387.
[15] T. Lammers, F. Kiessling, M. Ashford, W. Hennink, D. Crommelin, G. Storm, Cancer nanomedicine: is targeting our target? Nat. Rev. Mater. 1 (2016) 16069.
[16] V. Gutiérrez, A.B. Seabra, R.M. Reguera, J. Khandare, M. Calderón, New approaches from nanomedicine for treating leishmaniasis, Chem. Soc. Rev. 45 (2016) 152–168.
[17] K. Alamoudi, P. Martins, J.G. Croissant, S. Patil, H. Omar, N.M. Khashab, Thermo-responsive pegylated bubble liposome nanovectors for efficient siRNA delivery via endosomal escape, Nanomedicine 12 (2017) 1421–1433.
[18] H. Liu, Y. Yang, A. Wang, M. Han, W. Cui, J. Li, Hyperbranched polyglycerol-doped mesoporous silica nanoparticles for one- and two-photon activated photodynamic therapy, Adv. Funct. Mater. 26 (2016) 2561–2570.
[19] H. Li, K. Li, Y. Dai, X. Xu, X. Cao, Q. Zeng, H. He, L. Pang, J. Liang, X. Chen, In vivo near infrared fluorescence imaging and dynamic quantification of pancreatic metastatic tumors using folic acid conjugated biodegradable mesoporous silica nanoparticles, NanoMedicine 14 (2018) 1867–1877.
[20] Y. Lu, H. Fan, A. Stump, T.L. Ward, T. Rieker, C.J. Brinker, Aerosol-assisted self-assembly of mesostructured spherical nanoparticles, Nature 398 (1999) 223–226.
[21] Y. Lu, R. Ganguli, C.A. Drewien, M.T. Anderson, C.J. Brinker, W. Gong, Y. Guo, H. Soyez, B. Dunn, M.H. Huang, Continuous formation of supported cubic and hexagonal mesoporous films by sol–gel dip-coating, Nature 389 (1997) 364–368.
[22] C. Brinker, R. Kirkpatrick, D. Tallant, B. Bunker, B. Montez, NMR confirmation of strained "defects" in amorphous silica, J. Non Cryst. Solids 99 (1988) 418–428.
[23] M. Pfeiffer-Laplaud, D. Costa, F. Tielens, M.-P. Gaigeot, M. Sulpizi, Bimodal acidity at the amorphous silica/water interface, J. Phys. Chem. C 119 (2015) 27354–27362.
[24] A. Tsyganenko, E. Storozheva, O. Manoilova, T. Lesage, M. Daturi, J.-C. Lavalley, Brønsted acidity of silica silanol groups induced by adsorption of acids, Catal. Lett. 70 (2000) 159–163.
[25] R. De Farias, C. Airoldi, Thermogravimetry as a reliable tool to estimate the density of silanols on a silica gel surface, J. Therm. Anal. Calorim. 53 (1998) 751–756.
[26] S. Li, B.A. Moosa, J.G. Croissant, N.M. Khashab, Electrostatic assembly/disassembly of nanoscaled colloidosomes for light-triggered cargo release, Angew. Chem. Int. Ed. 127 (2015) 6908–6912.
[27] X. Cattoën, A. Noureddine, J. Croissant, N. Moitra, K. Bürglová, J. Hodačová, O. Cobos, M. Lejeune, F. Rossignol, D. Toulemon, S. Bégin-Colin, B. Pichon, L. Raehm, J.-O. Durand, M. Wong Chi Man, Click approaches in sol–gel chemistry, J. Sol-Gel Sci. Technol. 70 (2014) 245–253.
[28] A. Arkhireeva, J.N. Hay, Synthesis of sub-200 nm silsesquioxane particles using a modified Stöber sol–gel route, J. Mater. Chem. 13 (2003) 3122–3127.
[29] S. Li, Q. Wan, Z. Qin, Y. Fu, Y. Gu, Unraveling the mystery of Stöber silica's microporosity, Langmuir 32 (2016) 9180–9187.
[30] P.N. Durfee, Y.-S. Lin, D.R. Dunphy, A.J. Muñiz, K.S. Butler, K.R. Humphrey, A.J. Lokke, J.O. Agola, S.S. Chou, I.M. Chen, W. Wharton, J.L. Townson, C.L. Willman, C.J. Brinker, Mesoporous silica nanoparticle-supported lipid bilayers (protocells) for active targeting and delivery to individual leukemia cells, ACS Nano 10 (2016) 8325–8345.
[31] D. Tarn, C.E. Ashley, M. Xue, E.C. Carnes, J.I. Zink, C.J. Brinker, Mesoporous silica nanoparticle nanocarriers: biofunctionality and biocompatibility, Acc. Chem. Res. 46 (2013) 792–801.

[32] J. Croissant, X. Cattoën, M.W.C. Man, A. Gallud, L. Raehm, P. Trens, M. Maynadier, J.O. Durand, Biodegradable ethylene-bis (propyl) disulfide-based periodic mesoporous organosilica nanorods and nanospheres for efficient in-vitro drug delivery, Adv. Mater. 26 (2014) 6174–6180.
[33] J.G. Croissant, O. Mongin, V. Hugues, M. Blanchard-Desce, X. Cattoën, M. Wong Chi Man, V. Stojanovic, C. Charnay, M. Maynadier, M. Gary-Bobo, M. Garcia, L. Raehm, J.-O. Durand, Influence of the synthetic method on the properties of mesoporous silica nanoparticles designed for two-photon excitation, J. Mater. Chem. B 3 (2015) 5182–5188.
[34] J.G. Croissant, D. Zhang, S. Alsaiari, J. Lu, L. Deng, F. Tamanoi, A.M. AlMalik, J.I. Zink, N.M. Khashab, Protein–gold clusters-capped mesoporous silica nanoparticles for high drug loading, autonomous gemcitabine/doxorubicin co-delivery, and in-vivo tumor imaging, J. Control. Release 229 (2016) 183–191.
[35] J. Croissant, A. Chaix, O. Mongin, M. Wang, S. Clément, L. Raehm, J.-O. Durand, V. Hugues, M. Blanchard-Desce, M. Maynadier, A. Gallud, M. Gary-Bobo, M. Garcia, J. Lu, F. Tamanoi, D.P. Ferris, D. Tarn, J.I. Zink, Two-photon-triggered drug delivery via fluorescent nanovalves, Small 10 (2014) 1752–1755.
[36] J.G. Croissant, J.I. Zink, L. Raehm, J.O. Durand, Two-photon-excited silica and organosilica nanoparticles for spatiotemporal cancer treatment, Adv. Healthcare Mater. 7 (2018) 1701248.
[37] S. Dib, M. Boufatit, S. Chelouaou, F. Sadi-Hassaine, J. Croissant, J. Long, L. Raehm, C. Charnay, J.O. Durand, Versatile heavy metals removal via magnetic mesoporous nanocontainers, RSC Adv. 4 (2014) 24838–24841.
[38] J. Croissant, M. Maynadier, A. Gallud, H. Peindy N'Dongo, J.L. Nyalosaso, G. Derrien, C. Charnay, J.-O. Durand, L. Raehm, F. Serein-Spirau, N. Cheminet, T. Jarrosson, O. Mongin, M. Blanchard-Desce, M. Gary-Bobo, M. Garcia, J. Lu, F. Tamanoi, D. Tarn, T.M. Guardado-Alvarez, J.I. Zink, Two-photon-triggered drug delivery in cancer cells using nanoimpellers, Angew. Chem. Int. Ed. 125 (2013) 14058–14062.
[39] S. Giret, M. Wong Chi Man, C. Carcel, Mesoporous–silica-functionalized nanoparticles for drug delivery, Chem. Eur. J. 21 (2015) 13850–13865.
[40] N. Moitra, P. Trens, L. Raehm, J.-O. Durand, X. Cattoen, M.W.C. Man, Facile route to functionalized mesoporous silica nanoparticles by click chemistry, J. Mater. Chem. 21 (2011) 13476–13482.
[41] M. Gary-Bobo, Y. Mir, C. Rouxel, D. Brevet, I. Basile, M. Maynadier, O. Vaillant, O. Mongin, M. Blanchard-Desce, A. Morère, M. Garcia, J.-O. Durand, L. Raehm, Mannose-functionalized mesoporous silica nanoparticles for efficient two-photon photodynamic therapy of solid tumors, Angew. Chem. Int. Ed. 123 (2011) 11627–11631.
[42] E. Chelebaeva, L. Raehm, J.O. Durand, Y. Guari, J. Larionova, C. Guerin, A. Trifonov, M. Willinger, K. Thangavel, A. Lascialfari, O. Mongin, Y. Mir, M. Blanchard-Desce, Mesoporous silica nanoparticles combining two-photon excited fluorescence and magnetic properties, J. Mater. Chem. 20 (2010) 1877–1884.
[43] V. Lebret, L. Raehm, J.-O. Durand, M. Smaïhi, M.H.V. Werts, M. Blanchard-Desce, D. Méthy-Gonnod, C. Dubernet, Folic acid-targeted mesoporous silica nanoparticles for two-photon fluorescence, J. Biomed. Nanotechnol. 6 (2010) 176–180.
[44] D. Brevet, M. Gary-Bobo, L. Raehm, S. Richeter, O. Hocine, K. Amro, B. Loock, P. Couleaud, C. Frochot, A. Morere, P. Maillard, M. Garcia, J.O. Durand, Mannose-targeted mesoporous silica nanoparticles for photodynamic therapy, Chem. Commun. 12 (2009) 1475–1477.
[45] S.K. Alsaiari, M.A. Hammami, J.G. Croissant, H.W. Omar, P. Neelakanda, T. Yapici, K.-V. Peinemann, N.M. Khashab, Colloidal gold nanoclusters spiked silica fillers in

mixed matrix coatings: simultaneous detection and inhibition of healthcare-associated infections, Adv. Healthcare Mater. 6 (2017) 1601135.
[46] J.G. Croissant, C. Qi, M. Maynadier, X. Cattoën, M. Wong Chi Man, L. Raehm, O. Mongin, M. Blanchard-Desce, M. Garcia, M. Gary-Bobo, J.-O. Durand, Multifunctional gold-mesoporous silica nanocomposites for enhanced two-photon imaging and therapy of cancer cells, Front. Mol. Biosci. 3 (2016) 1, https://doi.org/10.3389/fmolb.2016.00001.
[47] J.G. Croissant, C. Qi, O. Mongin, V. Hugues, M. Blanchard-Desce, L. Raehm, X. Cattoen, M. Wong Chi Man, M. Maynadier, M. Gary-Bobo, M. Garcia, J.I. Zink, J.-O. Durand, Disulfide-gated mesoporous silica nanoparticles designed for two-photon-triggered drug release and imaging, J. Mater. Chem. B 3 (2015) 6456–6461.
[48] J. Croissant, J.I. Zink, Nanovalve-controlled cargo release activated by plasmonic heating, J. Am. Chem. Soc. 134 (2012) 7628–7631.
[49] Z. Li, J.C. Barnes, A. Bosoy, J.F. Stoddart, J.I. Zink, Mesoporous silica nanoparticles in biomedical applications, Chem. Soc. Rev. 41 (2012) 2590–2605.
[50] M.W. Ambrogio, C.R. Thomas, Y.-L. Zhao, J.I. Zink, J.F. Stoddartt, Mechanized silica nanoparticles: a new frontier in theranostic nanomedicine, Acc. Chem. Res. 44 (2011) 903–913.
[51] C. Wu, Y. Liang, K. Yang, Y. Min, Z. Liang, L. Zhang, Y. Zhang, Clickable periodic mesoporous organosilica monolith for highly efficient capillary chromatographic separation, Anal. Chem. 88 (2016) 1521–1525.
[52] A. Galarneau, J. Iapichella, D. Brunel, F. Fajula, Z. Bayram-Hahn, K. Unger, G. Puy, C. Demesmay, J.L. Rocca, Spherical ordered mesoporous silicas and silica monoliths as stationary phases for liquid chromatography, J. Sep. Sci. 29 (2006) 844–855.
[53] W.-H. Peng, Y.-Y. Lee, C. Wu, K.C.-W. Wu, Acid–base bi-functionalized, large-pored mesoporous silica nanoparticles for cooperative catalysis of one-pot cellulose-to-HMF conversion, J. Mater. Chem. 22 (2012) 23181–23185.
[54] X. Yang, D. Chen, S. Liao, H. Song, Y. Li, Z. Fu, Y. Su, High-performance Pd–Au bimetallic catalyst with mesoporous silica nanoparticles as support and its catalysis of cinnamaldehyde hydrogenation, J. Catal. 291 (2012) 36–43.
[55] H.T. Chen, B.G. Trewyn, J.W. Wiench, M. Pruski, V.S.Y. Lin, Urea and thiourea-functionalized mesoporous silica nanoparticle catalysts with enhanced catalytic activity for Diels–Alder reaction, Top. Catal. 53 (2010) 187–191.
[56] M. Ferre, R. Pleixats, M. Wong Chi Man, X. Cattoen, Recyclable organocatalysts based on hybrid silicas, Green Chem. 18 (2016) 881–922.
[57] M. Karimi, I. Chaudhury, C. Jianjun, M. Safari, R. Sadeghi, M. Habibi-Rezaei, J. Kokini, Immobilization of endo-inulinase on non-porous amino functionalized silica nanoparticles, J. Mol. Catal. B: Enzym. 104 (2014) 48–55.
[58] M. Hartmann, Ordered mesoporous materials for bioadsorption and biocatalysis, Chem. Mater. 17 (2005) 4577–4593.
[59] A. Vinu, K.Z. Hossain, K. Ariga, Recent advances in functionalization of mesoporous silica, J. Nanosci. Nanotechnol. 5 (2005) 347–371.
[60] D. Moelans, P. Cool, J. Baeyens, E.F. Vansant, Using mesoporous silica materials to immobilise biocatalysis-enzymes, Catal. Commun. 6 (2005) 307–311.
[61] J.L. Townson, Y.-S. Lin, J.O. Agola, E.C. Carnes, H.S. Leong, J.D. Lewis, C.L. Haynes, C.J. Brinker, Re-examining the size/charge paradigm: differing in vivo characteristics of size and charge-matched mesoporous silica nanoparticles, J. Am. Chem. Soc. 135 (2013) 16030.
[62] C. Pisani, J.-C. Gaillard, M. Odorico, J.L. Nyalosaso, C. Charnay, Y. Guari, J. Chopineau, J.-M. Devoisselle, J. Armengaud, O. Prat, The timeline of corona formation around silica nanocarriers highlights the role of the protein interactome, Nanoscale 9 (2017) 1840–1851.

[63] J.G. Croissant, X. Cattoën, J.-O. Durand, M.W.C. Man, N.M. Khashab, Organosilica hybrid nanomaterials with a high organic content: syntheses and applications of silsesquioxanes, Nanoscale 8 (2016) 19945–19972.
[64] J. Croissant, M. Maynadier, O. Mongin, V. Hugues, M. Blanchard-Desce, A. Chaix, X. Cattoën, M. Wong Chi Man, A. Gallud, M. Gary-Bobo, M. Garcia, L. Raehm, J.-O. Durand, Enhanced two-photon fluorescence imaging and therapy of cancer cells via gold@bridged silsesquioxane nanoparticles, Small 11 (2015) 295–299.
[65] Y. Fatieiev, J. Croissant, S. Alsaiari, B.A. Moosa, D.H. Anjum, N.M. Khashab, Photoresponsive bridged silsesquioxane nanoparticles with tunable morphology for light-triggered plasmid DNA delivery, ACS Appl. Mater. Interfaces 7 (2015) 24993–24997.
[66] J.G. Croissant, C. Mauriello-Jimenez, M. Maynadier, X. Cattoën, M.W.C. Man, L. Raehm, O. Mongin, M. Blanchard-Desce, M. Garcia, M. Gary-Bobo, Synthesis of disulfide-based biodegradable bridged silsesquioxane nanoparticles for two-photon imaging and therapy of cancer cells, Chem. Commun. 51 (2015) 12324–12327.
[67] Y. Fatieiev, J. Croissant, K. Julfakyan, L. Deng, D.H. Anjum, A. Gurinov, N. Khashab, Enzymatically degradable hybrid organic–inorganic bridged silsesquioxane nanoparticles for in vitro imaging, Nanoscale 7 (2015) 15046–15050.
[68] J.G. Croissant, X. Cattoën, M. Wong Chi Man, J.O. Durand, N.M. Khashab, Syntheses and applications of periodic mesoporous organosilica nanoparticles, Nanoscale 7 (2015) 20318–20334.
[69] C.M. Jimenez, D. Aggad, J.G. Croissant, K. Tresfield, D. Laurencin, D. Berthomieu, N. Cubedo, M. Rossel, S. Alsaiari, D.H. Anjum, R. Sougrat, M.A. Roldan-Gutierrez, S. Richeter, E. Oliviero, L. Raehm, C. Charnay, X. Cattoën, S. Clément, M.W.C. Man, M. Maynadier, V. Chaleix, V. Sol, M. Garcia, M. Gary-Bobo, N.M. Khashab, N. Bettache, J.-O. Durand, Porous porphyrin-based organosilica nanoparticles for nir two-photon photodynamic therapy and gene delivery in zebrafish, Adv. Funct. Mater. 28 (2018) 1800235.
[70] A. Noureddine, C.J. Brinker, Pendant/bridged/mesoporous silsesquioxane nanoparticles: versatile and biocompatible platforms for smart delivery of therapeutics, Chem. Eng. J. 340 (2018) 125–147.
[71] N. Mizoshita, T. Tani, S. Inagaki, Syntheses, properties and applications of periodic mesoporous organosilicas prepared from bridged organosilane precursors, Chem. Soc. Rev. 40 (2011) 789–800.
[72] B. Guan, Y. Cui, Z. Ren, Z.-a. Qiao, L. Wang, Y. Liu, Q. Huo, Highly ordered periodic mesoporous organosilica nanoparticles with controllable pore structures, Nanoscale 4 (2012) 6588–6596.
[73] D. Aggad, C.M. Jimenez, S. Dib, J.G. Croissant, L. Lichon, D. Laurencin, S. Richeter, M. Maynadier, S.K. Alsaiari, M. Boufatit, Gemcitabine delivery and photodynamic therapy in cancer cells via porphyrin-ethylene-based periodic mesoporous organosilica nanoparticles, ChemNanoMat 4 (2018) 46–51.
[74] J. Croissant, D. Salles, M. Maynadier, O. Mongin, V. Hugues, M. Blanchard-Desce, X. Cattoën, M. Wong Chi Man, A. Gallud, M. Garcia, M. Gary-Bobo, L. Raehm, J.-O. Durand, Mixed periodic mesoporous organosilica nanoparticles and core–shell systems, application to in vitro two-photon imaging, therapy and drug delivery, Chem. Mater. 26 (2014) 7214–7220.
[75] J. Ouwehand, J. Lauwaert, D. Esquivel, K. Hendrickx, V. Van Speybroeck, J.W. Thybaut, P. Van Der Voort, Facile synthesis of cooperative acid–base catalysts by clicking cysteine and cysteamine on an ethylene-bridged periodic mesoporous organosilica, Eur. J. Inorg. Chem. 2016 (2016) 2144–2151.
[76] E. De Canck, I. Ascoop, A. Sayari, P. Van Der Voort, Periodic mesoporous organosilicas functionalized with a wide variety of amines for CO_2 adsorption, Phys. Chem. Chem. Phys. 15 (2013) 9792–9799.

[77] C. Vercaemst, M. Ide, P.V. Wiper, J.T. Jones, Y.Z. Khimyak, F. Verpoort, P. Van Der Voort, Ethenylene-bridged periodic mesoporous organosilicas: from E to Z, Chem. Mater. 21 (2009) 5792–5800.
[78] J. Croissant, Y. Fatieiev, K. Julfakyan, J. Lu, A. Emwas, D. Anjum, H. Omar, F. Tamanoi, J. Zink, N. Khashab, Biodegradable oxamide-phenylene-based mesoporous organosilica nanoparticles with unprecedented drug payloads for delivery in cells, Chem. A Eur. J. 22 (2016) 14806–14811.
[79] A.C. Gomes, M.A. Lourenço, S.M. Bruno, P. Ferreira, A.A. Valente, M. Pillinger, I.S. Gonçalves, Post-synthetic modification of crystal-like periodic mesoporous phenylene-silica with ferrocenyl groups, J. Organomet. Chem. 751 (2014) 501–507.
[80] Y. Li, S. Wang, M. Xiao, M. Wang, Z. Huang, B. Li, Y. Yang, Chirality of the 1,4-phenylene–silica nanoribbons at the nano and angstrom levels, Nanotechnology 24 (2013), 035603.
[81] T. Okada, K.-i. Yamanaka, Y. Hirose, Y. Goto, T. Tani, S. Inagaki, Fluorescence studies on phenylene moieties embedded in a framework of periodic mesoporous organosilica, Phys. Chem. Chem. Phys. 13 (2011) 7961–7967.
[82] X. Liu, W. Zhuang, B. Li, L. Wu, S. Wang, Y. Li, Y. Yang, Helical periodic mesoporous 1,4-phenylene-silica nanorods with chiral crystalline walls, Chem. Commun. 47 (2011) 7215–7217.
[83] M. Ohashi, M.P. Kapoor, S. Inagaki, Chemical modification of crystal-like mesoporous phenylene-silica with amino group, Chem. Commun. 7 (2008) 841–843.
[84] Y. Li, A. Keilbach, M. Kienle, Y. Goto, S. Inagaki, P. Knochel, T. Bein, Hierarchically structured biphenylene-bridged periodic mesoporous organosilica, J. Mater. Chem. 21 (2011) 17338–17344.
[85] Y. Li, H. Wang, L. Wang, F. Zhou, Y. Chen, B. Li, Y. Yang, Handedness inversion in preparing chiral 4,4′-biphenylene-silica nanostructures, Nanotechnology 22 (2011) 135605.
[86] H. Li, B. Li, Y. Chen, M. Zhang, S. Wang, Y. Li, Y. Yang, Preparation of chiral 4,4′-biphenylene-silica nanoribbons, Chin. J. Chem. 27 (2009) 1860–1862.
[87] R. Bouchal, M. Daurat, M. Gary-Bobo, A. Da Silva, L. Lesaffre, D. Aggad, A. Godefroy, P. Dieudonné, C. Charnay, J.-O. Durand, Biocompatible periodic mesoporous ionosilica nanoparticles with ammonium walls: application to drug delivery, ACS Appl. Mater. Interfaces 9 (2017) 32018–32025.
[88] N. Ishito, K. Nakajima, Y. Maegawa, S. Inagaki, A. Fukuoka, Facile formation of gold nanoparticles on periodic mesoporous bipyridine-silica, Catal. Today 298 (2017) 258–262.
[89] M. Gao, S. Han, Y. Hu, J.J. Dynes, X. Liu, D. Wang, A pH-driven molecular shuttle based on rotaxane-bridged periodic mesoporous organosilicas with responsive release of guests, RSC Adv. 6 (2016) 27922–27932.
[90] X. Liu, Y. Maegawa, Y. Goto, K. Hara, S. Inagaki, Heterogeneous catalysis for water oxidation by an iridium complex immobilized on bipyridine-periodic mesoporous organosilica, Angew. Chem. Int. Ed. 128 (2016) 8075–8079.
[91] M. Waki, Y. Maegawa, K. Hara, Y. Goto, S. Shirai, Y. Yamada, N. Mizoshita, T. Tani, W.-J. Chun, S. Muratsugu, M. Tada, A. Fukuoka, S. Inagaki, A solid chelating ligand: periodic mesoporous organosilica containing 2,2′-bipyridine within the pore walls, J. Am. Chem. Soc. 136 (2014) 4003–4011.
[92] C. Mauriello-Jimenez, J. Croissant, M. Maynadier, X. Cattoen, M. Wong Chi Man, J. Vergnaud, V. Chaleix, V. Sol, M. Garcia, M. Gary-Bobo, L. Raehm, J.-O. Durand, Porphyrin-functionalized mesoporous organosilica nanoparticles for two-photon imaging of cancer cells and drug delivery, J. Mater. Chem. B 3 (2015) 3681–3684.
[93] P. Huang, X. Qian, Y. Chen, L. Yu, H. Lin, L. Wang, Y. Zhu, J. Shi, Metalloporphyrin-encapsulated biodegradable nanosystems for highly efficient magnetic

resonance imaging-guided sonodynamic cancer therapy, J. Am. Chem. Soc. 139 (2017) 1275–1284.
[94] C. Mauriello-Jimenez, M. Henry, D. Aggad, L. Raehm, C. Charnay, X. Cattoen, M.W.C. Man, S. Alpugan, V. Ahsen, D.K. Tarakcı, Porphyrin-or phthalocyanine-bridged silsesquioxane theranostic nanoparticles for two-photon imaging, photodynamic therapy or photoacoustic imaging, Nanoscale 9 (2017) 16622–16626.
[95] P. Van Der Voort, D. Esquivel, E. De Canck, F. Goethals, I. Van Driessche, F.J. Romero-Salguero, Periodic mesoporous organosilicas: from simple to complex bridges; a comprehensive overview of functions, morphologies and applications, Chem. Soc. Rev. 42 (2013) 3913–3955.
[96] C. Urata, H. Yamada, R. Wakabayashi, Y. Aoyama, S. Hirosawa, S. Arai, S. Takeoka, Y. Yamauchi, K. Kuroda, Aqueous colloidal mesoporous nanoparticles with ethenylene-bridged silsesquioxane frameworks, J. Am. Chem. Soc. 133 (2011) 8102–8105.
[97] S. Datz, H. Engelke, C.V. Schirnding, L. Nguyen, T. Bein, Lipid bilayer-coated curcumin-based mesoporous organosilica nanoparticles for cellular delivery, Microporous Mesoporous Mater. 225 (2016) 371–377.
[98] J.G. Croissant, Y. Fatieiev, H. Omar, D.H. Anjum, A. Gurinov, J. Lu, F. Tamanoi, J.I. Zink, N.M. Khashab, Periodic mesoporous organosilica nanoparticles with controlled morphologies and high drug/dye loadings for multicargo delivery in cancer cells, Chem. A Eur. J. 22 (2016) 9607–9615.
[99] T. Shimizu, K. Kanamori, A. Maeno, H. Kaji, C.M. Doherty, K. Nakanishi, Transparent ethenylene-bridged polymethylsiloxane aerogels: mechanical flexibility and strength and availability for addition reaction, Langmuir 33 (2017) 4543–4550.
[100] V. Rebbin, R. Schmidt, M. Fröba, Spherical particles of phenylene-bridged periodic mesoporous organosilica for high-performance liquid chromatography, Angew. Chem. Int. Ed. 45 (2006) 5210–5214.
[101] F. Hoffmann, M. Cornelius, J. Morell, M. Fröba, Silica-based mesoporous organic–inorganic hybrid materials, Angew. Chem. Int. Ed. Engl. 45 (2006) 3216–3251.
[102] F. Auras, Y. Li, F. Löbermann, M. Döblinger, J. Schuster, L.M. Peter, D. Trauner, T. Bein, A zinc phthalocyanine based periodic mesoporous organosilica exhibiting charge transfer to fullerenes, Chem. A Eur. J. 20 (2014) 14971–14975.
[103] H. Takeda, M. Ohashi, T. Tani, O. Ishitani, S. Inagaki, Enhanced photocatalysis of rhenium (I) complex by light-harvesting periodic mesoporous organosilica, Inorg. Chem. 49 (2010) 4554–4559.
[104] S. Inagaki, O. Ohtani, Y. Goto, K. Okamoto, M. Ikai, K.-i. Yamanaka, T. Tani, T. Okada, Light harvesting by a periodic mesoporous organosilica chromophore, Angew. Chem. Int. Ed. 48 (2009) 4042–4046.
[105] Y. Fatieiev, J. Croissant, K. Alamoudi, N. Khashab, Cellular internalization and biocompatibility of periodic mesoporous organosilica nanoparticles with tunable morphologies: from nanospheres to nanowires, ChemPlusChem 82 (2017) 631–637.
[106] M.A. Hammami, J.G. Croissant, L. Francis, S.K. Alsaiari, D.H. Anjum, N. Ghaffour, N.M. Khashab, Engineering hydrophobic organosilica nanoparticle-doped nanofibers for enhanced and fouling resistant membrane distillation, ACS Appl. Mater. Interfaces 9 (2017) 1737–1745.
[107] D.J. Conley, J.C. Carey, Biogeochemistry: silica cycling over geologic time, Nat. Geosci. 8 (2015) 431.
[108] P. Cermeño, P.G. Falkowski, O.E. Romero, M.F. Schaller, S.M. Vallina, Continental erosion and the cenozoic rise of marine diatoms, Proc. Natl. Acad. Sci. 112 (2015) 4239–4244.

[109] N.C. Marty, F. Claret, A. Lassin, J. Tremosa, P. Blanc, B. Madé, E. Giffaut, B. Cochepin, C. Tournassat, A database of dissolution and precipitation rates for clay-rocks minerals, Appl. Geochem. 55 (2015) 108–118.
[110] C. Ehlert, K. Doering, K. Wallmann, F. Scholz, S. Sommer, P. Grasse, S. Geilert, M. Frank, Stable silicon isotope signatures of marine pore waters—biogenic opal dissolution versus authigenic clay mineral formation, Geochim. Cosmochim. Acta 191 (2016) 102–117.
[111] S.W. Squyres, R.E. Arvidson, S. Ruff, R. Gellert, R. Morris, D. Ming, L. Crumpler, J. Farmer, D. Des Marais, A. Yen, Detection of silica-rich deposits on Mars, Science 320 (2008) 1063–1067.
[112] H.A. Currie, C.C. Perry, Silica in plants: biological, biochemical and chemical studies, Ann. Bot. 100 (2007) 1383–1389.
[113] P.J. Tréguer, C.L. De La Rocha, The world ocean silica cycle, Ann. Rev. Mar. Sci. 5 (2013) 477–501.
[114] P. Van Cappellen, S. Dixit, J. van Beusekom, Biogenic silica dissolution in the oceans: reconciling experimental and field-based dissolution rates, Global Biogeochem. Cycles 16 (4) (2002) 23–1.
[115] K. Pradeep, M. Nepolian, P. Anandhan, R. Kaviyarasan, M.V. Prasanna, S. Chidambaram, in: A study on variation in dissolved silica concentration in groundwater of hard rock aquifers in southeast coast of India, IOP Conference Series: Materials Science and Engineering, IOP Publishing, 2016, p. 012008.
[116] D.S. Lawson, D.C. Hurd, H.S. Pankratz, Silica dissolution rates of decomposing phytoplankton assemblages at various temperatures, Am. J. Sci. 278 (1978) 1373–1393.
[117] K.D. Bidle, F. Azam, Accelerated dissolution of diatom silica by marine bacterial assemblages, Nature 397 (1999) 508–512.
[118] H. Ehrlich, K.D. Demadis, O.S. Pokrovsky, P.G. Koutsoukos, Modern views on desilicification: biosilica and abiotic silica dissolution in natural and artificial environments, Chem. Rev. 110 (2010) 4656–4689.
[119] K. Yamashita, H. Daiguji, Molecular dynamics simulations of water uptake into a silica nanopore, J. Phys. Chem. C 119 (2015) 3012–3023.
[120] J.L. Paris, M. Colilla, I. Izquierdo-Barba, M. Manzano, M. Vallet-Regí, Tuning mesoporous silica dissolution in physiological environments: a review, J. Mater. Sci. 52 (2017) 8761–8771.
[121] W. Vogelsberger, J. Schmidt, Dissolution of commercial nanoscale silica particles in electrolyte solution: the importance of the solid-solvent-ratio to physical and chemical properties of the solid–liquid interface, Colloids Surf. A Physicochem. Eng. Asp. 537 (2018) 591–603.
[122] K. Braun, A. Pochert, M. Beck, R. Fiedler, J. Gruber, M. Lindén, Dissolution kinetics of mesoporous silica nanoparticles in different simulated body fluids, J. Sol-Gel Sci. Technol. 79 (2016) 319–327.
[123] H. Yamada, C. Urata, Y. Aoyama, S. Osada, Y. Yamauchi, K. Kuroda, Preparation of colloidal mesoporous silica nanoparticles with different diameters and their unique degradation behavior in static aqueous systems, Chem. Mater. 24 (2012) 1462–1471.
[124] V. Cauda, A. Schlossbauer, T. Bein, Bio-degradation study of colloidal mesoporous silica nanoparticles: effect of surface functionalization with organo-silanes and poly (ethylene glycol), Microporous Mesoporous Mater. 132 (2010) 60–71.
[125] D. Cassano, D. Rota Martir, G. Signore, V. Piazza, V. Voliani, Biodegradable hollow silica nanospheres containing gold nanoparticle arrays, Chem. Commun. 51 (2015) 9939–9941.
[126] D. Shen, J. Yang, X. Li, L. Zhou, R. Zhang, W. Li, L. Chen, R. Wang, F. Zhang, D. Zhao, Biphase stratification approach to three-dimensional dendritic biodegradable mesoporous silica nanospheres, Nano Lett. 14 (2014) 923–932.

[127] Q. He, J. Shi, M. Zhu, Y. Chen, F. Chen, The three-stage in vitro degradation behavior of mesoporous silica in simulated body fluid, Microporous Mesoporous Mater. 131 (2010) 314–320.
[128] K.S. Finnie, D.J. Waller, F.L. Perret, A.M. Krause-Heuer, H.Q. Lin, J.V. Hanna, C.J. Barbé, Biodegradability of sol–gel silica microparticles for drug delivery, J. Sol-Gel Sci. Technol. 49 (2009) 12–18.
[129] P.J. Kempen, S. Greasley, K.A. Parker, J.L. Campbell, H.-Y. Chang, J.R. Jones, R. Sinclair, S.S. Gambhir, J.V. Jokerst, Theranostic mesoporous silica nanoparticles biodegrade after pro-survival drug delivery and ultrasound/magnetic resonance imaging of stem cells, Theranostics 5 (2015) 631.
[130] G. Chen, Z. Teng, X. Su, Y. Liu, G. Lu, Unique biological degradation behavior of Stöber mesoporous silica nanoparticles from their interiors to their exteriors, J. Biomed. Nanotechnol. 11 (2015) 722–729.
[131] Y. Li, J. Shi, Hollow-structured mesoporous materials: chemical synthesis, functionalization and applications, Adv. Mater. 26 (2014) 3176–3205.
[132] R.K. Iler, The Chemistry of Silica: Solubility, Polymerization, Colloid and Surface Properties, and Biochemistry, Wiley, 1979.
[133] F. Turci, C. Pavan, R. Leinardi, M. Tomatis, L. Pastero, D. Garry, S. Anguissola, D. Lison, B. Fubini, Revisiting the paradigm of silica pathogenicity with synthetic quartz crystals: the role of crystallinity and surface disorder, Part. Fibre Toxicol. 13 (2016) 32.
[134] K.D. Demadis, E. Mavredaki, Green additives to enhance silica dissolution during water treatment, Environ. Chem. Lett. 3 (2005) 127–131.
[135] P.B. McMahon, D.A. Vroblesky, P.M. Bradley, F.H. Chapelle, C.D. Gullett, Evidence for enhanced mineral dissolution in organic acid-rich shallow ground water, Groundwater 33 (1995) 207–216.
[136] J. Depasse, A. Watillon, The stability of amorphous colloidal silica, J. Colloid Interface Sci. 33 (1970) 430–438.
[137] T.W. Healy, Stability of Aqueous Silica Sols, ACS Publications, 1994.
[138] S. Seré, B. De Roo, M. Vervaele, S. Van Gool, S. Jacobs, J.W. Seo, J.-P. Locquet, Altering the biodegradation of mesoporous silica nanoparticles by means of experimental parameters and surface functionalization, J. Nanomater. 2018 (2018), 7390618.
[139] K. Chen, J. Zhang, H. Gu, Dissolution from inside: a unique degradation behaviour of core–shell magnetic mesoporous silica nanoparticles and the effect of polyethyleneimine coating, J. Mater. Chem. 22 (2012) 22005–22012.
[140] Y.-S. Lin, N. Abadeer, K.R. Hurley, C.L. Haynes, Ultrastable, redispersible, small, and highly organomodified mesoporous silica nanotherapeutics, J. Am. Chem. Soc. 133 (2011) 20444–20457.
[141] V. Cauda, C. Argyo, T. Bein, Impact of different PEGylation patterns on the long-term bio-stability of colloidal mesoporous silica nanoparticles, J. Mater. Chem. 20 (2010) 8693–8699.
[142] E. Choi, S. Kim, How can doxorubicin loading orchestrate in vitro degradation behaviors of mesoporous silica nanoparticles under a physiological condition? Langmuir 33 (2017) 4974–4980.
[143] E. von Haartman, D. Lindberg, N. Prabhakar, J.M. Rosenholm, On the intracellular release mechanism of hydrophobic cargo and its relation to the biodegradation behavior of mesoporous silica nanocarriers, Eur. J. Pharm. Sci. 95 (2016) 17–27.
[144] Y. Yang, Y. Niu, J. Zhang, A.K. Meka, H. Zhang, C. Xu, C.X.C. Lin, M. Yu, C. Yu, Biphasic synthesis of large-pore and well-dispersed benzene bridged mesoporous organosilica nanoparticles for intracellular protein delivery, Small 11 (2015) 2743–2749.

[145] M. Zhou, X. Du, W. Li, X. Li, H. Huang, Q. Liao, B. Shi, X. Zhang, M. Zhang, One-pot synthesis of redox-triggered biodegradable hybrid nanocapsules with a disulfide-bridged silsesquioxane framework for promising drug delivery, J. Mater. Chem. B 5 (23) (2017) 4455–4469.
[146] Z. Xu, K. Zhang, X. Liu, H. Zhang, A new strategy to prepare glutathione responsive silica nanoparticles, RSC Adv. 3 (2013) 17700–17702.
[147] L. Sun, D. Wang, Y. Chen, L. Wang, P. Huang, Y. Li, Z. Liu, H. Yao, J. Shi, Core–shell hierarchical mesostructured silica nanoparticles for gene/chemo-synergetic stepwise therapy of multidrug-resistant cancer, Biomaterials 133 (2017) 219–228.
[148] X. Du, F. Kleitz, X. Li, H. Huang, X. Zhang, S.Z. Qiao, Disulfide-bridged organosilica frameworks: designed, synthesis, redox-triggered biodegradation, and nanobiomedical applications, Adv. Funct. Mater. 28 (2018) 1707325.
[149] K.M. Rao, S. Parambadath, A. Kumar, C.-S. Ha, S.S. Han, Tunable intracellular degradable periodic mesoporous organosilica hybrid nanoparticles for doxorubicin drug delivery in cancer cells, ACS Biomater Sci. Eng. 4 (2017) 175–183.
[150] L. Maggini, I. Cabrera, A. Ruiz-Carretero, E.A. Prasetyanto, E. Robinet, L. De Cola, Breakable mesoporous silica nanoparticles for targeted drug delivery, Nanoscale 8 (2016) 7240–7247.
[151] E.A. Prasetyanto, A. Bertucci, D. Septiadi, R. Corradini, P. Castro-Hartmann, L. De Cola, Breakable hybrid organosilica nanocapsules for protein delivery, Angew. Chem. Int. Ed. 55 (2015) 3323–3327.
[152] L. Qiu, Y. Zhao, B. Li, Z. Wang, L. Cao, L. Sun, Triple-stimuli (protease/redox/pH) sensitive porous silica nanocarriers for drug delivery, Sensors Actuators B Chem. 240 (2017) 1066–1074.
[153] S. Quignard, S. Masse, G. Laurent, T. Coradin, Introduction of disulfide bridges within silica nanoparticles to control their intra-cellular degradation, Chem. Commun. 49 (2013) 3410–3412.
[154] Y. Chen, Q. Meng, M. Wu, S. Wang, P. Xu, H. Chen, Y. Li, L. Zhang, L. Wang, J. Shi, Hollow mesoporous organosilica nanoparticles: a generic intelligent framework-hybridization approach for biomedicine, J. Am. Chem. Soc. 136 (2014) 16326–16334.
[155] Y. Chen, J. Shi, Chemistry of mesoporous organosilica in nanotechnology: molecularly organic–inorganic hybridization into frameworks, Adv. Mater. 28 (2016) 3235–3272.
[156] M. Wu, Q. Meng, Y. Chen, Y. Du, L. Zhang, Y. Li, L. Zhang, J. Shi, Large-pore ultrasmall mesoporous organosilica nanoparticles: micelle/precursor co-templating assembly and nuclear-targeted gene delivery, Adv. Mater. 27 (2015) 215–222.
[157] Z. Teng, X. Su, B. Lee, C. Huang, Y. Liu, S. Wang, J. Wu, P. Xu, J. Sun, D. Shen, Yolk–shell structured mesoporous nanoparticles with thioether-bridged organosilica frameworks, Chem. Mater. 26 (2014) 5980–5987.
[158] N. Lu, Y. Tian, W. Tian, P. Huang, Y. Liu, Y. Tang, C. Wang, S. Wang, Y. Su, Y. Zhang, J. Pan, Z. Teng, G. Lu, Smart cancer cell targeting imaging and drug delivery system by systematically engineering periodic mesoporous organosilica nanoparticles, ACS Appl. Mater. Interfaces 8 (2016) 2985–2993.
[159] N. Lu, W. Fan, X. Yi, S. Wang, Z. Wang, R. Tian, O. Jacobson, Y. Liu, B.C. Yung, G. Zhang, Biodegradable hollow mesoporous organosilica nanotheranostics for mild hyperthermia-induced bubble-enhanced oxygen-sensitized radiotherapy, ACS Nano 12 (2018) 1580–1591.
[160] L. Zhang, L. Wang, H. Yao, F. Xu, Y. Chen, Biodegradable and biocompatible monodispersed hollow mesoporous organosilica with large pores for delivering biomacromolecules, J. Mater. Chem. B 5 (2017) 8013–8025.
[161] J.D. Rocca, M.E. Werner, S.A. Kramer, R.C. Huxford-Phillips, R. Sukumar, N.D. Cummings, J.L. Vivero-Escoto, A.Z. Wang, W. Lin, Polysilsesquioxane

nanoparticles for triggered release of cisplatin and effective cancer chemoradiotherapy, Nanomedicine 11 (2015) 31–38.
[162] J.D. Rocca, R.C. Huxford, E. Comstock-Duggan, W. Lin, Polysilsesquioxane nanoparticles for targeted platin-based cancer chemotherapy by triggered release, Angew. Chem. Int. Ed. 50 (2011) 10330–10334.
[163] J. Yue, S.Z. Luo, M.M. Lu, D. Shao, Z. Wang, W.F. Dong, A comparison of mesoporous silica nanoparticles and mesoporous organosilica nanoparticles as drug vehicles for cancer therapy, Chem. Biol. Drug Des. 92 (2) (2018) 1435–1444(ahead of print).
[164] S.P.H. Moghaddam, M. Yazdimamaghani, H. Ghandehari, Glutathione-sensitive hollow mesoporous silica nanoparticles for controlled drug delivery, J. Control. Release 282 (2018) 62–75(ahead of print).
[165] B.T. Vu, S.A. Shahin, J.G. Croissant, Y. Fatieviev, K. Matsumoto, T.L. Hoang Doan, T. Yik, A. Conteras, L. Ratliff, C. Mauriello Jimenez, L. Raehm, N.M. Khashab, J.-O. Durand, C. Glackin, F. Tamanoi, Chick chorioallantoic membrane assay as an in vivo model to study the effect of nanoparticle-based anticancer drugs in ovarian cancer, Sci. Rep. 8 (1) (2018), 8524(ahead of print).
[166] J.L. Vivero-Escoto, D. DeCillis, L. Fritts, D.L. Vega, in: Porphyrin-based polysilsesquioxane nanoparticles to improve photodynamic therapy for cancer treatment, Proc. SPIE 8931, Optical Methods for Tumor Treatment and Detection: Mechanisms and Techniques in Photodynamic Therapy XXIII, 89310Z, 2014.
[167] J.L. Vivero-Escoto, W.J. Rieter, H. Lau, R.C. Huxford-Phillips, W. Lin, Biodegradable polysilsesquioxane nanoparticles as efficient contrast agents for magnetic resonance imaging, Small 9 (2013) 3523–3531.
[168] N. Lu, P. Huang, W. Fan, Z. Wang, Y. Liu, S. Wang, G. Zhang, J. Hu, W. Liu, G. Niu, Tri-stimuli-responsive biodegradable theranostics for mild hyperthermia enhanced chemotherapy, Biomaterials 126 (2017) 39–48.
[169] L. Yu, Y. Chen, M. Wu, X. Cai, H. Yao, L. Zhang, H. Chen, J. Shi, "Manganese extraction" strategy enables tumor-sensitive biodegradability and theranostics of nanoparticles, J. Am. Chem. Soc. 138 (2016) 9881–9894.
[170] X. Hao, X. Hu, C. Zhang, S. Chen, Z. Li, X. Yang, H. Liu, G. Jia, D. Liu, K. Ge, Hybrid mesoporous silica-based drug carrier nanostructures with improved degradability by hydroxyapatite, ACS Nano 9 (2015) 9614–9625.
[171] X. Li, L. Zhang, X. Dong, J. Liang, J. Shi, Preparation of mesoporous calcium doped silica spheres with narrow size dispersion and their drug loading and degradation behavior, Microporous Mesoporous Mater. 102 (2007) 151–158.
[172] A. Sousa, K. Souza, E. Sousa, Mesoporous silica/apatite nanocomposite: special synthesis route to control local drug delivery, Acta Biomater. 4 (2008) 671–679.
[173] M. Musa, T.P. Kannan, I. Ab Rahman, Assessment of DNA damage caused by locally produced hydroxyapatite–silica nanocomposite using comet assay on human lung fibroblast cell line, Mol. Cell. Toxicol. 8 (2012) 53–60.
[174] S. Latifi, M. Fathi, M. Golozar, Preparation and characterisation of bioactive hydroxyapatite–silica composite nanopowders via sol–gel method for medical applications, Mol. Cell. Toxicol. 110 (2011) 8–14.
[175] X. Ge, C. Li, C. Fan, X. Feng, B. Cao, Enhanced photoluminescence properties of methylene blue dye encapsulated in nanosized hydroxyapatite/silica particles with core–shell structure, Appl. Phys. A 113 (2013) 583–589.
[176] P.A. Williamson, P.J. Blower, M.A. Green, Synthesis of porous hollow silica nanostructures using hydroxyapatite nanoparticle templates, Chem. Commun. 47 (2011) 1568–1570.
[177] Y. He, B. Zeng, S. Liang, M. Long, H. Xu, Synthesis of pH-responsive biodegradable mesoporous silica–calcium phosphate hybrid nanoparticles as a high potential drug carrier, ACS Appl. Mater. Interfaces 9 (2017) 44402–44409.
[178] G. Ma, X.Y. Liu, Hydroxyapatite: hexagonal or monoclinic? Cryst. Growth Des. 9 (2009) 2991–2994.

[179] A. Llopis-Lorente, B. Lozano-Torres, A. Bernardos, R. Martínez-Máñez, F. Sancenón, Mesoporous silica materials for controlled delivery based on enzymes, J. Mater. Chem. B 5 (17) (2017) 3069–3083.
[180] J. Liu, B. Zhang, Z. Luo, X. Ding, J. Li, L. Dai, J. Zhou, X. Zhao, J. Ye, K. Cai, Enzyme responsive mesoporous silica nanoparticles for targeted tumor therapy in vitro and in vivo, Nanoscale 7 (2015) 3614–3626.
[181] Y.-J. Cheng, G.-F. Luo, J.-Y. Zhu, X.-D. Xu, X. Zeng, D.-B. Cheng, Y.-M. Li, Y. Wu, X.-Z. Zhang, R.-X. Zhuo, Enzyme-induced and tumor-targeted drug delivery system based on multifunctional mesoporous silica nanoparticles, ACS Appl. Mater. Interfaces 7 (2015) 9078–9087.
[182] Y.-L. Sun, Y. Zhou, Q.-L. Li, Y.-W. Yang, Enzyme-responsive supramolecular nanovalves crafted by mesoporous silica nanoparticles and choline-sulfonatocalix[4]arene[2]pseudorotaxanes for controlled cargo release, Chem. Commun. 49 (2013) 9033–9035.
[183] L. Deng, G. Wang, J. Ren, B. Zhang, J. Yan, W. Li, N.M. Khashab, Enzymatically triggered multifunctional delivery system based on hyaluronic acid micelles, RSC Adv. 2 (2012) 12909–12914.
[184] C. Coll, L. Mondragon, R. Martinez-Manez, F. Sancenon, M. Dolores Marcos, J. Soto, P. Amoros, E. Perez-Paya, Enzyme-mediated controlled release systems by anchoring peptide sequences on mesoporous silica supports, Angew. Chem. Int. Ed. 50 (2011) 2138–2140.
[185] A. Bernardos, L. Mondragon, E. Aznar, M.D. Marcos, R. Martinez-Manez, F. Sancenon, J. Soto, J.M. Barat, E. Perez-Paya, C. Guillem, P. Amoros, Enzyme-responsive intracellular controlled release using nanometric silica mesoporous supports capped with "saccharides" ACS Nano 4 (2010) 6353–6368.
[186] N.H. Tsao, E.A.H. Hall, Enzyme-degradable hybrid polymer/silica microbubbles as ultrasound contrast agents, Langmuir 32 (2016) 6534–6543.
[187] A. Corma, U. Díaz, M. Arrica, E. Fernández, Í. Ortega, Organic–inorganic nanospheres with responsive molecular gates for drug storage and release, Angew. Chem. Int. Ed. 121 (2009) 6365–6368.
[188] B. Rühle, P. Saint-Cricq, J.I. Zink, Externally controlled nanomachines on mesoporous silica nanoparticles for biomedical applications, ChemPhysChem 17 (2016) 1769–1779.
[189] B.Y. Lee, Z. Li, D.L. Clemens, B.J. Dillon, A.A. Hwang, J.I. Zink, M.A. Horwitz, Redox-triggered release of moxifloxacin from mesoporous silica nanoparticles functionalized with disulfide snap-tops enhances efficacy against pneumonic tularemia in mice, Small 12 (2016) 3690–3702.
[190] D. Tarn, D.P. Ferris, J.C. Barnes, M.W. Ambrogio, J.F. Stoddart, J.I. Zink, A reversible light-operated nanovalve on mesoporous silica nanoparticles, Nanoscale 6 (2014) 3335–3343.
[191] L. Maggini, L. Travaglini, I. Cabrera, P. Castro-Hartmann, L. De Cola, Biodegradable peptide–silica nanodonuts, Chem. A Eur. J. 22 (2016) 3697–3703.
[192] L. Wang, M. Huo, Y. Chen, J. Shi, Iron-engineered mesoporous silica nanocatalyst with biodegradable and catalytic framework for tumor-specific therapy, Biomaterials 163 (2018) 1–13.
[193] K.K. Pohaku Mitchell, A. Liberman, A.C. Kummel, W.C. Trogler, Iron(III)-doped, silica nanoshells: a biodegradable form of silica, J. Am. Chem. Soc. 134 (2012) 13997–14003.
[194] Y.-K. Peng, Y.-J. Tseng, C.-L. Liu, S.-W. Chou, Y.-W. Chen, S.E. Tsang, P.-T. Chou, One-step synthesis of degradable T_1-FeOOH functionalized hollow mesoporous silica nanocomposites from mesoporous silica spheres, Nanoscale 7 (2015) 2676–2687.
[195] Y.-K. Peng, C.-L. Liu, H.-C. Chen, S.-W. Chou, W.-H. Tseng, Y.-J. Tseng, C.-C. Kang, J.-K. Hsiao, P.-T. Chou, Antiferromagnetic iron nanocolloids: a new

generation in vivo T1 MRI contrast agent, J. Am. Chem. Soc. 135 (2013) 18621–18628.
[196] K.P. Mitchell, S. Sandoval, M. Cortes-Mateos, J. Alfaro, A. Kummel, W. Trogler, Self-assembled targeting of cancer cells by iron (III)-doped, silica nanoparticles, J. Mater. Chem. B 2 (2014) 8017–8025.
[197] H. Omar, J.G. Croissant, K. Alamoudi, S. Alsaiari, I. Alradwan, M.A. Majrashi, D.H. Anjum, P. Martins, B. Moosa, A. Almalik, N.M. Khashab, Biodegradable magnetic silica@iron oxide nanovectors with ultra-large mesopores for high protein loading, magnetothermal release, and delivery, J. Control. Release 259 (2017) 187–194.

FSC
www.fsc.org
MIX
Papier | Fördert
gute Waldnutzung
FSC® C083411

Kolibri 360 GmbH
Ferdinand-Jühlke-Straße 7
99095 Erfurt, Deutschland
produktsicherheit@kolibri360.de